Technische Universität Dresden

# Hierarchical Resource Allocation for Powerline Communications (PLC) Networks

**Le Phu Do**

der Fakultät Elektrotechnik und Informationstechnik der Technischen Universität Dresden

zur Erlangung des akademischen Grades eines

**Doktoringenieurs**

(Dr.-Ing.)

genehmigte Dissertation

Vorsitzender:

Prof. Dr.-Ing. Peter Schegner (TU Dresden)

Gutachter:

Prof. Dr.-Ing. Ralf Lehnert (TU Dresden)

Prof. Dr. Andrea M. Tonello (DIEGM – University of Udine, Italien)

Tag der Einreichung: 02.04.2012

Tag der Verteidigung: 18.09.2012

**Bibliografische Information der Deutschen Nationalbibliothek**
Die Deutsche Nationalbibliothek verzeichnet diese Publikation in der Deutschen Nationalbibliografie; detaillierte bibliografische Daten sind im Internet über
**http://dnb.d-nb.de** abrufbar.

ISBN 978-3-8439-0718-7


Sternstr. 18, 80538 München
Tel.: 089/66060798
www.dr.hut-verlag.de



1. Auflage 2012

Technische Universität Dresden

# Hierarchical Resource Allocation for Powerline Communications (PLC) Networks

**Le Phu Do**

Department of Electrical Engineering and Information Technology
Technische Universität Dresden

approved dissertation to obtain the academic degree

**Doktoringenieur**

(Dr.-Ing.)

Chairman:

Prof. Dr.-Ing. Peter Schegner (TU Dresden)

Reviewers:

Prof. Dr.-Ing. Ralf Lehnert (TU Dresden)

Prof. Dr. Andrea M. Tonello (DIEGM – University of Udine, Italien)

Date of submission: 02.04.2012

Date of defense: 18.09.2012

## Abstract

Powerline Communications (PLC) is a technology which utilizes the available electrical power network to provide telecommunication services to the customer network terminals. A PLC network is planned and installed in an area to provide telecommunication services to the users, who are connected to these network terminals. In accordance with the coverage and the actual size of the electrical power lines' infrastructure in that area, the entire PLC network may be divided into multiple, smaller PLC clusters. In this work, it is assumed that in each PLC network cluster, called a PLC cell, one communications base station (BS) is installed to serve all the connected subscribers as a master-slave system. In a PLC cell, all the network devices share the same physical transmission medium.

In case a PLC network consists of several adjacent cells, in which parallel transmissions in different cells may interfere, a cell should not be allowed to utilize the whole transmission resources. To each cell should only be allocated a subset of the available transmission resources. In order to achieve this, the transmission resources are partitioned into a set of the so-called inter-cell resource units, or channels, to be allocated to each cell. These allocated channels are reserved for the transmission inside that cell. Also the channel conditions may be time variant.

In a PLC cell, the users do not move, but their traffic demands change during run-time. This causes the active users and traffic demands in each cell to vary. The overall multi-cell PLC structure may also be dynamically changed by the activation or deactivation of some cells. If each cell has a constant number of channels, users in different cells may receive a different amount of resources for their transmissions during run-time. This situation leads to potential unfairness between users in different cells. Therefore, a method which allows the dynamic shifting of the transmission resources between cells is required. Compared to the cellular mobile wireless networks, a distributed resource allocation scheme is necessary, because it is not always possible to organize a central controller for the PLC cells that may belong to different service providers or private owners.

The problem of sharing transmission resources, not only between users inside a cell but also between users in different cells, has to be solved. The resource allocation issue is divided into two sub-problems, which form the so-called two-level hierarchical resource allocation model. Separating the entire resource allocation problem into two sub-problems matches the PLC network model in which the relationships between PLC network devices are logically organized on two levels: between PLC cells (inter-cell level) and between users in a PLC cell (intra-cell level). A BS in a PLC cell is the network device, which performs the resource organization for the two sides: one side receives the reserved resources for its cell and the other manages and organizes the allocated resources for its users.

For inter-PLC-cell resource allocation, a Distributed Inter-Cell Resource Allocation (DICRA) protocol is proposed. This protocol allows the cells to cooperate and dynamically share channels in a distributed fashion. Neighboring PLC cells communicate locally with each other in order to negotiate and organize the transmission resources. Employing this algorithm, a cell exchanges its channel allocation information with its neighbors via the Channel Announcement Messages (CA-Msg). Each cell organizes its own resources to achieve the highest number of reserved channels while avoiding possible interference with neighboring cells. If two cells do not interfere with each other, a channel can be used by both cells.

The negotiation strategy is based on the following two procedures: seizing free channels and releasing channels in case a cell has more channels per active user than its neighbors. The protocol achieves a high channel reuse factor, maintains fairness between users in different cells in terms of the number of channels per active user, and avoids usage of the same channels in neighboring cells. An active user is the user who has data to transmit to or receive from its master. Adaptation time and stability of the protocol are also analyzed. The performance of the protocol strongly depends on the number of neighbors of a cell, but not on the size of the PLC network

A certain number of transmission resource units are required by each cell to meet the traffic requirements inside that cell. Based on the cell structure, the amount of resources requested by a specific service by a particular user is calculated. In case the number of reserved resource units is lower than the amount of resources available for the requested traffic, the resource allocation protocols on the intra-cell level have to determine which services to guarantee. Once accepted, a service is given enough resources in accordance with its Quality of Service (QoS) requirements.

For intra-PLC-cell traffic scheduling, the resource allocation problem is formulated as an optimization problem. The objectives are maximizing resource utilization and maintaining fairness between users under the constraint of the available resources. This optimization problem is solved by different methods, namely Exhaustive Search (ES), Round Robin Scheduling (RRS), simulated annealing based Heuristic Algorithm (HA), and the proposed Benefit Based Selection (BBS) algorithm. An adaptation scheme, which allows using knowledge from the previous scheduling for the current search, is proposed to improve the calculation time when applying the HA algorithm. The performances of the scheduling methods are evaluated based on the Time Division Multiple Access (TDMA) based MAC layer for different referenced PLC cells.

The scheduling problem is also formulated for the combination of the Orthogonal Frequency Division Multiple Access (OFDMA) and the TDMA and solved by an extension of the BBS (called EBBS). Network utilization using the OFDMA/TDMA is higher than with the TDMA when the channels have different data rates, however, at the cost of computation time.

## Zusammenfassung

Powerline Communications (PLC) ist eine Technologie, die das verfügbare Stromnetz nutzt um Kundenanschlüsse mit Telekommunikationsdiensten zu versorgen. Ein PLC Netzwerk wird in einem Gebiet geplant und errichtet um den Nutzern mit Stromanschluss diese Telekommunikationsdienste zur Verfügung zu stellen. Abhängig von der eigentlichen Größe und Abdeckung der Stromversorgungs-infrastruktur in einem Gebiet, kann das geplante PLC Netzwerk auch in mehrere kleine PLC Cluster unterteilt werden. Im Zusammenhang dieser Arbeit wird angenommen, dass in jedem PLC Netzwerkcluster, auch PLC Zelle genannt, eine zentrale Basisstation (BS) alle angeschlossenen Teilnehmer entsprechend eines Master-Slave-Sytems bedient. In einem solchen PLC Netzwerk teilen sich alle Netzwerkgeräte das selbe physikalischen Medium um auf das Netzwerk zuzugreifen.

Im Falle eines PLC Netzwerks, das aus mehreren, aneinander grenzenden Zellen besteht, könnte eine Zelle die anderen beeinträchtigen. Deswegen sollte es einer Zelle nicht erlaubt sein die gesamten Übertragungsresourcen für die Übertragungen ihrer Netzwerkgeräte zu nutzen. Jede Zelle darf nur einen Teil der verfügbaren Übertragungskapazität benutzen. Um das zu erreichen wird die Übertragungskapazität in Gruppen von sogenannten interzellulären Resourceneinheiten, oder Kanäle, eingeteilt und diese werden von den einzelnen Zellen genutzt. Die Kanäle werden für die Übertragung innerhalb einer Zelle reserviert. Zudem können die Kanalcharakterisiken zeitvariant sein.

In PLC Zellen bewegen sich die Nutzer zwar nicht, aber der von ihnen verursachte Datenverkehr ändert sich über der Zeit. Dies führt zu stetig variierenden Verkehrsaufkommen und Nutzerzahlen in jeder Zelle. Die mehrere Zellen umfassende PLC Infrastruktur könnte auch durch das Abschalten und Hinzuschalten einiger Zellen dynamisch gestaltet werden. Wenn jede Zelle aber eine gleichbleibende Anzahl an Kanälen besitzt, könnten Nutzern unterschiedlicher Zellen unterschiedlich viele Resourcen für ihre Übertragungen zur Verfügung stehen. Das führt zur Benachteiligung von Nutzern in einigen Zellen. Deswegen wird eine Methode zur dynamischen Verlagerung von Übertragungskapazitäten zwischen den Zellen benötigt. Im Gegensatz zu einem zellularen Mobilfunknetz, ist in PLC eine verteilte Resourcenzuteilung notwendig. Eine zentrale Steuerung aller Zellen ist nicht immer möglich, da diese zu verschiedenen Dienstanbietern oder auch privaten Besitzer gehören können.

Das Problem der Aufteilung von Übertragungskapazitäten, nicht nur zwischen den Nutzern einer Zelle, sondern auch zwischen Nutzern verschiedener Zellen, muss gelöst werden. Die Resourcenzuteilung wird dabei in zwei Teilprobleme unterteilt, die ein sogenanntes zweistufiges Übertragungsresourcen-Zuteilungsmodell bilden. Dabei entsprechen die zwei Teilprobleme dem PLC Netzwerkmodell, wo die Beziehungen zwischen PLC Netzwerkgeräten in zwei logischen Stufen organisiert werden: zwischen PLC Zellen (inter-cell level) und zwischen Nutzern einer PLC Zelle (intra-cell level). Jede BS in einer PLC Zelle übernimmt die Resourceneinteilung auf zwei Seiten: eine Seite empfängt die reservierten Resourcen für diese Zelle und die Andere organisiert deren Verteilung auf die Nutzer.

Zur Zuordnung der Übertragungsresourcen wird das sogenannte Distributed Inter-Cell Resource Allocation (DICRA) Protokoll vorgeschlagen. Dieses Protokoll ermöglicht es den Zellen zusammenzuarbeiten und die Kanäle dynamisch und auf dezentralisierte Weise zuzuteilen. Benachbarte PLC Zellen kommunizieren dabei untereinander um über die Übertragungsresourcen zu verhandeln und diese zu verwalten. Bei der Anwendung dieses Algorithmus tauscht eine Zelle die Informationen der Kanalzuteilung mit ihren Nachbarn mittels einer sogenannten Channel Announcement Message (CA-Msg) aus und verteilt die eigenen Resourcen so, dass die höchstmöglische Anzahl an reservierten Kanälen erreicht wird ohne die benachbarten Zellen zu beeinträchtigen. Wenn zwei Zellen einander nicht beeinträchtigen, kann derselbe Kanal von beiden Zellen genutzt werden.

Die Verhandlungsstrategie basiert auf den zwei folgenden Prozeduren: freie Kanäle werden reserviert und Kanäle werden wieder freigegeben, wenn eine Zelle mehr Kanäle je aktivem Nutzer hat als ihre Nachbarn. Das Protokoll erreicht einen hohen Wiederverwendungsanteil (Reuse factor) bei den Kanälen, erhält dabei aber auch die Gleichberechtigung unter den Nutzern verschiedener Zellen in Bezug auf die Anzahl an Kanälen je aktivem Nutzer und es vermeidet die Nutzung gleicher Kanäle von benachbarten Zellen. Ein aktiver Nutzer ist hierbei ein Nutzer, der Daten an seinen Master schicken möchte oder diese empfängt. Es werden sowohl die Anpassungszeit als auch die Stabilität des Protokolls untersucht. Die Leistungsfähigkeit des Protokolls hängt stark von der Anzahl der Nachbarn einer bestimmten Zelle ab, aber nicht von der Größe des PLC Netzwerks insgesamt.

Jede Zelle benötigt eine bestimmte Menge von Einheiten der Übertragungsresourcen um die Verkehrsanforderungen in der Zelle zu erfüllen. Die Resourcenanforderungen werden anhand der Zellstruktur und der Anforderungen bestimmter Dienste der einzelnen Nutzer berechnet. Sollte die Anzahl an reservierten Resourcen unter den Resourcenanforderungen der Dienste liegen, dann muss das Protokoll für die Resourcenzuteilung innerhalb der Zelle die Menge an Übertragungsresourcen für die unterschiedlichen Quality-of-Service-Anforderungen (QoS) der laufenden Dienste bestimmen.

Für die Resourcenzuteilung innerhalb einer PLC Zelle wird das Problem der Resourcenzuteilung als ein Optimierungsproblem formuliert, wobei die Ziele eine maximale Resourcenausnutzung und der Erhalt der Gleichberechtigung (Fairness) unter den Nutzern anhand der verfügbaren Resourcen sind. Dieses Optimierungsproblem wird auf verschiedene Weisen gelöst, namentlich durch Exhaustive Search (ES), Round Robin Scheduling (RRS), simulated annealing based Heuristic Algorithm (HA) und der hier vorgestellte Benefit Based

Selection (BBS) Algorithmus. Eine Anpassungsmaßnahme für den HA Algorithmus wird vorgeschlagen, die durch Nutzung von Erfahrungen aus vorherigen Planungen die Berechnungszeit der aktuellen Planung verkürzt. Die Leistungsfähigkeit der verschiedenen Planungsmethoden wird mit Hilfe des Time Division Multiple Access (TDMA) nutzenden MAC Layers für unterschiedliche PLC Zellen bewertet.

Das Planungsproblem wird auch für eine Kombination aus Orthogonal Frequency Division Multiple Access (OFDMA) und TDMA formuliert und mit einer Erweiterung des BBS Algorithmus (EBBS) gelöst. Die Netzwerkauslastung ist mit OFDMA/TDMA besser als mit TDMA, wenn die Kanäle unterschiedliche Übertragungsraten haben, jedoch geht das auf Kosten der Rechenzeit.

## Acknowledgments

It is a pleasure to thank many people who helped and supported me during the work in this thesis. In the first place, I would like to express my sincere gratitude to my PhD supervisor Prof. Dr.-Ing Ralf Lehnert who gives me the opportunity to complete my doctoral study at the Chair for Telecommunications, Technische Universität Dresden. I would like to thank him for his support, advice, encouragement, and contribution to the development of this thesis. I would also like to thank Prof. Dr. Andrea M. Tonello from DIEGM – University of Udine, Italy for his review of this thesis.

I would like to thank the staffs, past and present, of the Chair for Telecommunications. I would like to specially thank Dr. Halid Hrasnica and Dr. Abdelfatteh Haidine for their expertise in the research area of PLC. Thanks to Dr. Qin Dai for sharing ideas from her quality of experience researches, Dr. Samer Sulaiman for his reliable support, Dr. Matthias Baumann and Dr. Roland Schingnitz for their helpful discussions about the theoretical approach.

I would like to thank all the participating partners in the European funded OPERA project for the investigations and ideas that are achieved during cooperation work. The knowledge in this thesis is also drawn from several other cooperation projects with other academic and industrial partners. I would like to thank Dr. Gerd Bumiller for his knowledge in PLC channel when he was in iAd GmbH. I would also like to thank Mr. Marco Dietrich and Mr. Hendrik Schwuchow from ELCON Systemtechnik GmbH for their practical expertise and valuable support.

Last but not least, I would like to thank my parents, brother, and my wife Van Anh Do with my sons Duc Phong Do and Duc Huy Do for their loving support, patience, and encouragement during all these years.

# Table of Contents

# Chapter 1 Introduction

## 1.1 Introduction to Powerline Communications

With the rapid increase in the demand of telecommunication services, the use of alternative technologies to improve transmission capacity is required. One of the technologies able to provide communication services to the end user is Powerline Communications technology (PLC). The PLC technology utilizes the existing electricity network as a transmission medium avoiding the necessity of constructing a new infrastructure. Nowadays, the power line network covers wide areas and supplies electric energy to almost every household. For this reason, the power grids are a promising infrastructure and can be found almost everywhere to provide network access to the users.

Generally, the PLC technique is used in two application domains: access networks and inhome communications. The access networks are used by service providers to provide network access to customers/subscribers. Based on the customers' locations and the size of the power line infrastructure, a PLC network may be divided into multiple PLC cells. In a PLC network, all network devices share the same physical medium to transmit their data. The head-end of a PLC network may be installed in a transformer station or a street cabinet in the electricity grid, and the customer network terminals are installed in front of the electricity meters. The inhome applications are used by the users privately inside their households. An inhome PLC network is constructed by several PLC modems plugged into the electricity lines to provide telecommunication services for equipment inside a household.

## 1.2 Motivations and Research Challenges

Despite the attraction of the potentially low cost of a PLC network by utilizing the available power lines, the realization of this kind of communications technique comes with several difficulties. From the point of view of the network design, the transmission medium is shared by all subscribers and the transmission resources are limited. Potential collisions of data packets could not only occur in different pieces of equipment within a PLC cell, but also between the equipment of adjacent cells.

In a situation, in which an access PLC cell is connected to a house containing an inhome PLC cell, the PLC communication signals can propagate through both cells. These adjacent cells are neighbors. Therefore, simultaneous transmissions in neighboring PLC cells in the same frequency band should be avoided. To do so, an inter-PLC-cell resource allocation protocol is required. The transmission resources are subdivided into smaller portions that are allocated to

PLC cells. Some coexistence protocols have been proposed for sharing the transmission resources between the PLC cells in a network. They are, however, not designed for optimizing the transmission capacity and do not allow for dynamically shifting resources between the PLC cells.

Dynamic resource allocation is necessary, because inside a PLC cell, users do not move but their traffic demands vary during run-time. Also the conditions of the transmission channels vary during run-time, thus making the scheduling problem more complex. Because the number of active users and their traffic demands in each PLC cell varies, the users in different PLC cells with fixed resources may be allocated different amounts of resources for their transmissions. This situation leads to the problem of unfairness between users in different cells. Entire multi-cell PLC network structures may also change dynamically due to the activation and deactivation of some cells, which also requires a method for re-arranging the unused resources among the working cells. Moreover, a distributed approach is required, as it is not possible to organize a central controller for the PLC cells that may belong to different service providers, or in the case of shared resources between access and inhome cells.

Inside a PLC cell, typically, a base station controls the entire medium accesses of its slaves through a master-slave model. That is why a scheduling scheme is needed for controlling the access to the available resources inside a PLC cell. Since PLC supports multiple users with multiple services, it requires a scheduler to manage the available transmission resources. This scheduler assigns access periods to users based on their requests and has to guarantee QoS for each type of service in accordance with the service requirements, e.g. QoS requirements are characterized by voice, video, web-browsing, data. In a system utilizing the TDMA access scheme, the resources that the base station assigns to each user are the time-slot duration for accessing the network.

The problem of the scheduler arises when there are not enough transmission resources available to satisfy all the requests from several users simultaneously. Therefore, scheduling strategies are proposed to find an effective way to distribute limited resources to users and requested services. The problem is formulated as an optimization problem and solved by different algorithms, including a heuristic algorithm. A so-called adaptive scheme is proposed to speed up the execution times of the heuristic algorithm. The scheduling problem is also solved with the combination of the OFDMA and TDMA access schemes.

## 1.3 Outline of the Thesis

The thesis is organized as follows: chapter 2 introduces the powerline communications with different technologies applied. In this chapter, functionalities of the PHY- and MAC-layers in PLC networks are reviewed. The chapter focuses on the transmission technology based on the

OPERA specification. Chapter 3 reviews possible scheduling techniques to reveal their potential application in PLC systems. By applying scheduling techniques, PLC networks allow the network providers to effectively utilize transmission resources to provide communication service to the end users with particular quality of service guarantee requirements. In chapter 4, the reference models used for this study are presented. In chapter 5, the so-called hierarchical resource allocation model is proposed. In accordance with this model, the transmission resources are firstly shared by the PLC cells. Subsequently, the allocated resources in each PLC cell are further divided into smaller portions for each user. Chapter 6 proposes a distributed dynamic resource allocation protocol for dynamically sharing resources between the PLC cells during run-time, based on the dynamic resource requirements from different cells. Chapter 7 formulates the scheduling problem as an optimization problem and solves it by means of various methods. In the first part of this chapter, the problem is analysed and solved with the TDMA access scheme. In the second part, the combination of the OFDMA and TDMA, which allows utilization of the best channels qualities from different users to improve the resource utilization, is investigated. Finally, chapter 8 concludes the main achievements of this thesis and outlines possible topics for future work.

# Chapter 2 Powerline Communications

## 2.1 Introduction to Powerline Communications

Powerline Communications (PLC) is a technology, which utilizes the available electricity network as transmission medium to provide telecommunication services to the users through network terminals. There are two application areas of PLC, namely access network and inhome application [Dost01][HrasHa04]. The access PLC network is used to provide communication services to the customers located in their household and the inhome PLC network is normally used by the customers themselves to provide connections between their communications equipment.

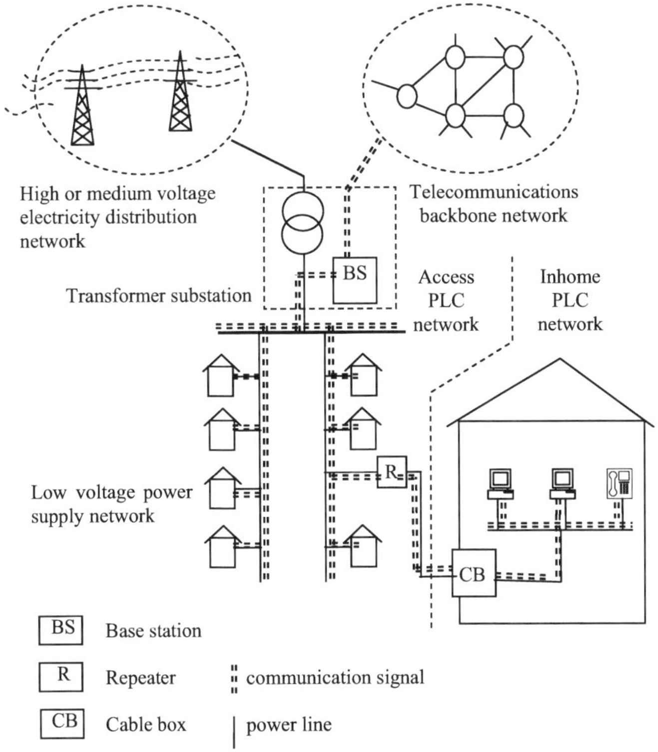


*Figure 2-1: PLC network over the power grid*

Fig. 2-1 presents a PLC network in an electricity grid. The network consists of both access and inhome PLC segments. The access PLC network is built upon low-voltage power lines. A base station (BS) is installed in the transformer station to provide network access to the subscribers in their households and connects the access network to the telecommunications backbone network. Since several customers connect to the base station through the same power cable, an access scheme is implemented in the base station to allow customers sharing of the same transmission medium. If the distance between a BS and the user is larger than the direct transmission range, repeaters are installed to increase the network coverage. In a private house, PLC equipment is installed to set up a private inhome PLC network. The PLC network allows the communications equipment to communicate over the existing power lines installed in the house.

In order to inject and extract the communication signal from the power line, particular PLC couplers are necessary. There are two kinds of coupling methods, namely capacitive coupler and inductive coupler, based on the component in the coupling module. Several coupling designs and their performances have been analyzed, e.g. as presented in [Dost01], [FezzRe06]. Fig. 2-2 presents the position of the coupling units in the transmission over the power line between a transmitter and a receiver. On the transmitter side, the coupling module injects the communication signal into the power line, in which the data is modulated into high frequency carriers. The signal propagates through the power line to the receiver side. On the receiver side, the communication signal is extracted by the coupling unit before forwarding it to the receiver.

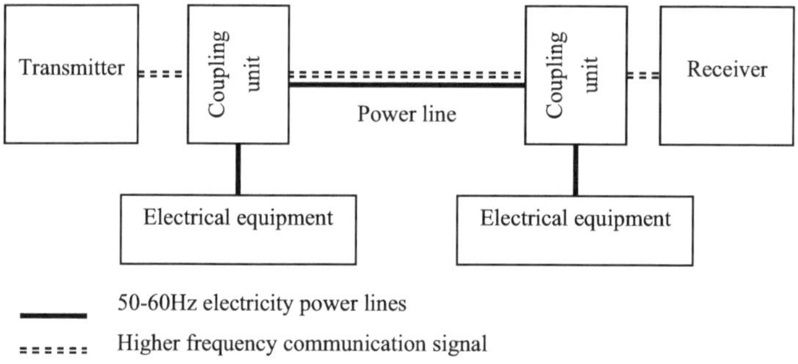


*Figure 2-2: Principle of Communication over Power Line Channel*

For the access PLC network, a coupler is installed in the BS to inject and split high frequency signals to and from the power line. Each customer is connected to the PLC access network via a PLC coupler, which is installed in front of his electricity meter. In the inhome PLC network, each PLC network terminal is equipped with a corresponding coupling unit.

## 2.2 PLC Systems

There are two main categories of PLC systems based on the used frequency band and the transmission data rate: broadband PLC (BB-PLC) and the narrowband PLC (NB-PLC). Due to the Electro-Magnetic Compatibility (EMC) requirements in different region regulations, only certain frequency bands and limited power emission can be used in the power lines. CENELEC permits NB-PLC transmissions in the frequency band between 9 kHz and 148.5 kHz, and the FCC (Federal Communications Commission) permits it in a band up to 500 kHz, where PLC can achieve a data rate of up to 500 kbps [D51RE03]. The BB- PLC, which utilizes the frequency band from 1 MHz to 30 MHz and above, typically transmits with a data rate of up to 200 Mbps [HomePl06] [Opera06]. Tab. 2-1 presents the classification of the different PLC systems, their potential applications as well as corresponding standards and specifications.

| | Frequency band | Transmission data rate | Standards/ Specifications | Applications |
|---|---|---|---|---|
| Narrowband PLC | CENELEC: 9kHz-148.5kHz/ FCC:9kHz-500kHz | Up to 150kbps/ 500kbps | ITU G.hnem, PRIME, G3, REMPLI, IEEE P1901.2 | SmartGrids, Automatic Meter Reading, Energy management, Command and Control |
| Broadband PLC | 1MHz-30MHz, Inhome: up to 50MHz | up to 200Mbps, Inhome up to 500Mbps | OPERA, UPA, HomePlug, HD-PLC, ITU G.hn, IEEE P1901 | Last mile access network, Multimedia, Inhome networking |

*Table 2-1: PLC system classification*

NB-PLC is suitable for applications requiring low data rates like signaling, switching of lamps, and automatic meter reading. This is utilized where real-time applications are not required. On the other hand, BB-PLC, with the ability to provide high-speed transmission, can be used for applications that require real-time transmission or high data rate like video transmissions, and internet access. Both BB- and NB-PLC are used in access networks and inhome applications.

### 2.2.1 PHY Layer for PLC

#### 2.2.1.1 Channel, Modulation, and Coding

The physical layer (PHY) provides the method of transmitting a data stream over the power line between end systems. Since the power line is not originally designed for transmissions of communication signals, transmission in the power line are affected by different noise types.

The noises caused by several sources, e.g. from electricity equipment and radio systems. In accordance with [Dost01], four types of noises affect the transmission in the power line: colored background noise, narrow-band interference, asynchronous impulse noise, and synchronous impulse noise. Fig. 2-3 presents the modeling of a PLC channel between a transmitter and a receiver. Detailed characteristics of the noise sources are given in [Dost01].

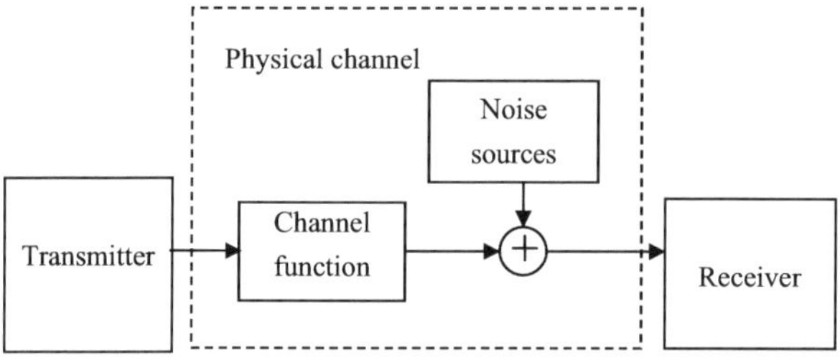


*Figure 2-3: Noise sources in a channel model*

Based on the frequency ranges and application requirements, several modulation schemes are applied to both NB- and BB-PLC. Fig. 2-4 presents some standard modulation schemes that are commonly utilized by PLC techniques.

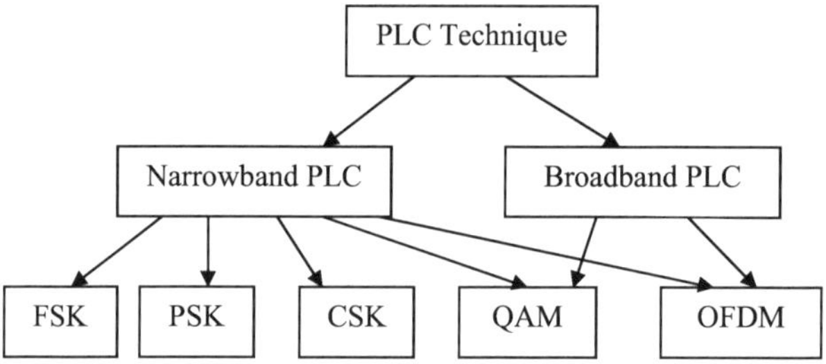


FSK: Frequency Shift Keying
PSK: Phase Shift Keying
CSK: Code Shift Keying
QAM: Quadrature Amplitude Modulation
OFDM: Orthogonal Frequency Division Multiplexing

*Figure 2-4: PLC techniques and modulation schemes*

Nowadays, most of the PLC systems utilize the Orthogonal Frequency Division Multiplexing (OFDM) modulation with bit-loading mechanisms for a flexible adaptation to the noisy transmission medium in the power lines.

#### 2.2.1.2 OFDM-Modulation

Orthogonal Frequency Division Multiplexing (OFDM) is a modulation technique that uses multiple carriers to transmit data. Different from the frequency division multiplexing (FDM) is that each carrier is orthogonal to the other carriers [BahaBu99]. Fig. 2-5 presents the spectrum of an OFDM signal. The OFDM provides higher efficiency in frequency exploitation, and therefore, it allows for the realization of higher data rates. The OFDM modulation scheme is used in several wired and wireless communication systems, e.g. in digital subscriber line (DSL), television and radio broadcasting services, and in power line communications (PLC).

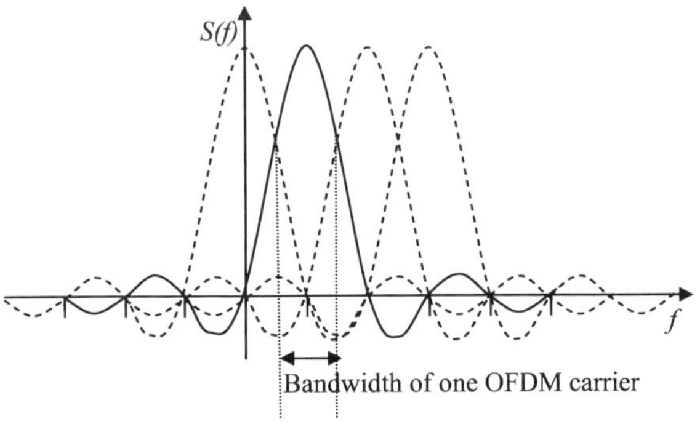


*Figure 2-5: Spectrum of an OFDM signal*

Utilizing the OFDM technique, the transmission overcomes the problem of multi-path delay spread, which causes the so-called inter symbol interference (ISI). This ISI decreases the performance of systems based on spread spectrum scheme. The OFDM symbol is significantly longer than the delayed paths, e.g. 71µs compared to a delay spread from 1µs up to 10µs for inhome and access PLC channels, and with the utilization of a guard interval, e.g. 5 µs, there is no ISI [Dort01].

At the transmitter, the serial data stream will be converted into $Ns$ parallel streams. Each OFDM carrier, $f_1, f_2 .. f_{Ns}$, carries one of that parallel streams to the receiver. At the receiver side, all received parallel streams are combined again to produce the original data flow. A simple flow-chart of the OFDM system is shown in Fig. 2-6. The IFFT converts the bits in an OFDM symbol into a time domain OFDM signal and the FFT converts the received signal back.

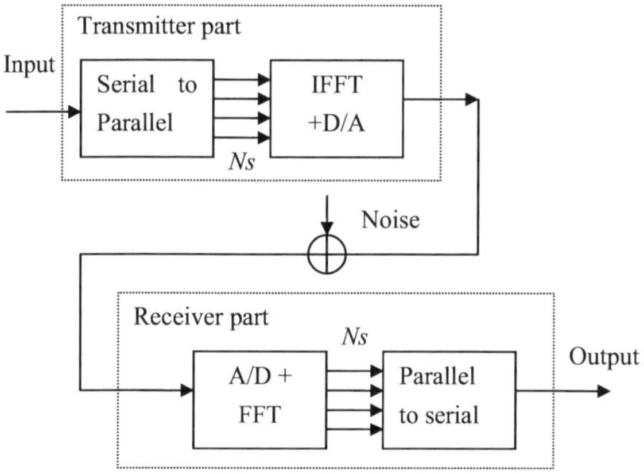


IFFT: Inverse Fast Fourier Transform A/D: Analog to Digital converter
FFT: Fast Fourier Transform D/A: Digital to Analog converter

*Figure 2-6: OFDM transmission system*

Another advantage of OFDM technology is that it allows the system to selectively turn on and off certain carriers. The PLC systems therefore can meet the local regulation requirement demanding to notch particular carriers to avoid interference with other systems using the same frequency bands, e.g. amateur radio and other wireless services. This is called "notching".

Two PLC channel measurement resulting in two different connections, taken from [D15OP07], are depicted in Fig. 2-7. They were created during a field test of an OPERA based broadband PLC access network over a low-voltage grid in Madrid. Each graph shows the distribution of the number of bits per carrier (BPC) for each carrier index during one hour. The results presented with two connections indicate that the channel of the connection in Fig. 2-7a is relatively stable, while the other (Fig. 2-7a) shows higher variation during the measurement period.

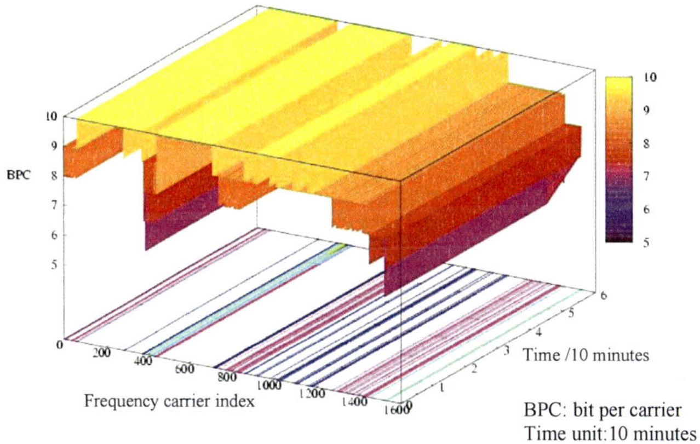


a) Channel with less variation during run-time

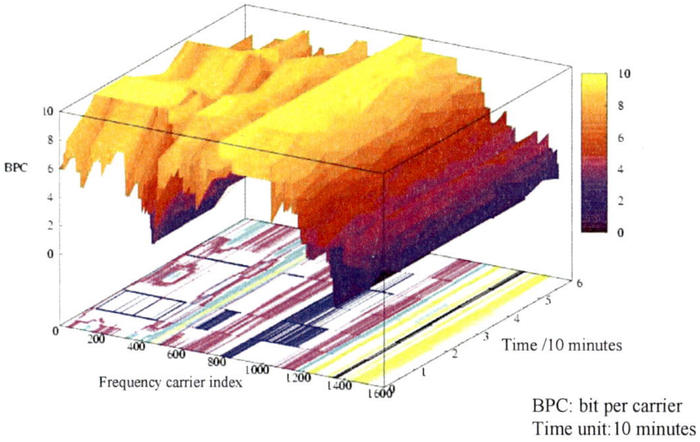


a) Channel with more variation during run-time

*Figure 2-7: PLC channel measurement results [source: D15OP07]*

## 2.2.2 MAC Layer for PLC

### 2.2.2.1 Multiple Access Scheme

The multiple access control (MAC) scheme divides the transmission resources into smaller portions to allow several users to access the network. Each user occupies a separate resource segment to transmit their data. As presented in [HrasHa04], three multiple access schemes and their combinations have been applied in the communications area.

In the time division multiple access scheme (TDMA), the entire frequency band is used by a user in a certain time slot. Usually, several time slots are organized into a time frame, so that each user is assigned a specific time slot in a time frame for its transmission.

The frequency division multiple access scheme (FDMA) divides the available frequency spectrum into smaller bands, each assigned to a particular user. When the OFDM modulation method is used, a band is organized as a group of carriers. The user occupies the assigned frequency band during the time of the transmission. The frequency bands can be re-assigned to other users and services based on the traffic requirement.

Another multiple access scheme is the code division multiple access (CDMA). This scheme uses the entire frequency spectrum and time frame for the transmissions of several users simultaneously. A receiver obtains the superposition of all signals and has to filter the signals by using a specific code sequence.

### 2.2.2.2 MAC Protocol Classification

Depending on the requirements, one or a combination of several MAC mechanisms can be used for the PLC system. A suitable MAC protocol improves the utilization of the limited transmission resources in the PLC network. Several MAC protocols are investigated for PLC networks, as presented in [Hras04] and [HrasHa04]. The MAC protocols can be classified as presented in Fig. 2-8.

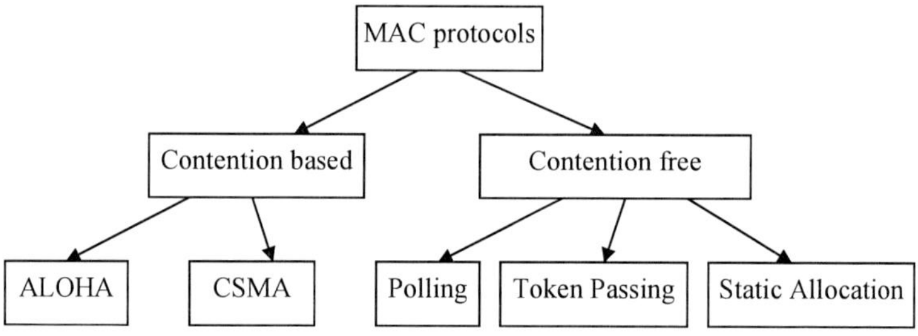


*Figure 2-8: Classification of the MAC protocols*

The ALOHA based protocols (pure ALOHA, slotted ALOHA) may be used for signaling in the PLC systems, in which the stations have to randomly send short messages. The carrier sense multiple access (CSMA, CSMA/CA) is effectively used with long messages. Nevertheless, the collision detection, which is required by the CSMA/CD, is not realizable in the PLC technique, due to the hidden terminal problem. Combined with corresponding collision resolution mechanisms, the network utilization can be improved when applying the contention protocols.

The contention free protocols like polling and token passing are suitable for systems, in which highly reliable communications are required, e.g. for automatic meter reading (AMR). The static allocation, e.g. of time slots and frequency bands to a specific user, is used when fixed transmission resources for users are necessary.

Based on the applications, a PLC network may have a specific controlling architecture. The nodes in a network may be configured to operate individually (e.g. using a contention based access scheme), under control of a master station as a master-slave model (e.g. using a contention free access scheme), or a hybrid of these methods (e.g. by switching between two models).

### 2.2.3 Coexistence

In accordance with application requirements, both the BB- and the NB-PLC can be used simultaneously without interfering with each other. This is because the distance between the frequency bands used is large enough. However, using more than one BB- or NB-PLC network requires a coexistence protocol, irrespective of whether the PLC networks are using the same or different technologies.

In [HomePa07], a so-called Inter PHY Protocol (IPP) has been proposed. Utilizing this protocol requires each PLC network to regularly broadcast its presence in a special message. When a PLC network detects the availability of its neighbors, it has to adjust its transmission resources in accordance with the number of neighbors. If there is one neighbor detected, the PLC network only uses half of the transmission resources. If it detects two neighboring cells, this cell has to reduce the used transmission resources to a third, and so on for additional neighbors. The protocol proposed in [D28OP07] use same principle.

## 2.3 The OPERA Specification

The investigations in this thesis use the parameters of the OPERA Specification [Opera06]. The network and transmission organization as well as the efficiency of the MAC protocol are reviewed in this section. Some important additional procedures are presented in Appendix A.1.5. This specification, published in January 2006, provides proposals for broadband-PLC including the technology and system descriptions. It is the result of the EU funded FP6 Open PLC European Research Alliance (OPERA) project. In the OPERA specification, functionalities of the PHY and MAC layers are specified. In the physical layer, this system utilizes the Orthogonal Frequency Division Multiplexing (OFDM) modulation technique. The MAC in the OPERA specification is a TDMA based protocol and employs a token to grant access rights to the network elements being a variant of the classic polling protocol.

### 2.3.1 Network Organization

There are three types of terminals to organize a PLC network: the base station (BS), the repeater (RP), and customer premises equipment (user). Repeaters are installed to increase the network coverage in case the distance between a BS and its user is larger than the direct transmission range. One example of a logical PLC network is presented in Fig. 2-9.

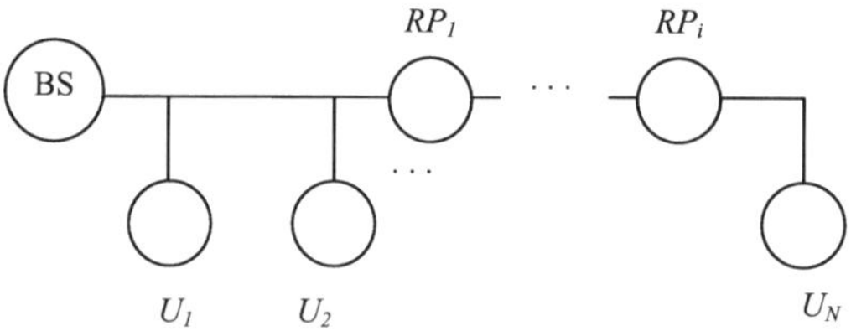


*Figure 2-9: Logical structure of a PLC network*

In accordance with the OPERA specification, a PLC network operates in a master-slave model in which all transmissions in the network are controlled by the BS. The MAC protocol in the BS uses the so-called token to deliver the access rights to users and repeaters. The BS generates a token and sends it along with the data to a user, who is granted to send in the next time period. The user transmits its data together with the token back to the BS in accordance with the information in the received token. Important MAC parameters and transmission sequences in a PLC network corresponding with the OPERA specification are presented in Appendix A.1.

### 2.3.2 Transmission Organization

To support real-time services, that request short network access intervals, like voice over Internet protocol (VoIP), the BS has to organize several transmissions in a time frame (TF). Thus, it requires several overhead units. On the contrary, for elastic traffic, e.g. Email or file transferring, the transmission may only get one time slot per TF. Therefore, it requires only a minimum overhead. By utilizing eight sub time frames (sTF) inside each TF, a service may be organized to transmit every $8^{th}$ sTF, every $4^{th}$ sTF, every $2^{nd}$ sTF or every sTF, which results in 1, 2, 4, or 8 times access to the network in a time frame. In Fig. 2-10, an organization for a

VoIP (from user $U_1$, access interval of 30ms), a web-traffic ($U_2$, access interval of 120ms), and a data service ($U_3$, access interval 240ms) is demonstrated. The allocated transmission duration for each user in both downlink (from BS to users) and uplink (from users to BS) transmission periods includes all necessary overhead for a transmission, e.g. signaling and transmission/reception switching time.

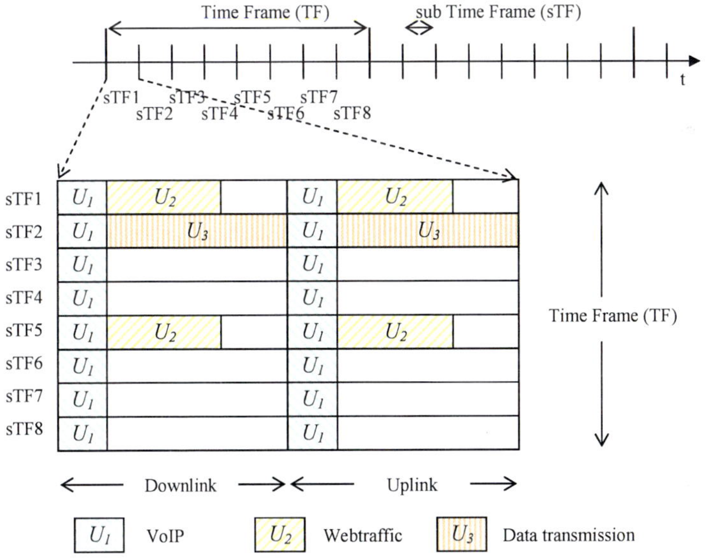


*Figure 2-10: Organization of transmissions in a time frame*

# Chapter 3 Scheduling Techniques and QoS Support for PLC Systems

This chapter presents the scheduling techniques and analyzes them to exploit their potential applications for PLC systems. The available schedulers that could be utilized in a PLC system have to be adapted to the particular characteristics of the PLC system in terms of resource utilization, fairness between multiple users, and fairness between multiple traffic flows of services by regarding QoS requirements. The PLC systems have to ensure sufficient QoS requirements for different types of telecommunication services.

The limited transmission resources have to be allocated to each user in accordance with a suitable resource allocation method. The QoS parameters for different kinds of services, and the survey of the previous studies concentrating on resource allocation as well as their applications to the PLC system are presented. The issues and the challenges that a PLC system experiences in order to ensure QoS support are also analyzed. The resource allocation is also analyzed for the networks in which more than one service provider delivers telecommunication services to the users in the same physical network. For large networks in which a cellular model is applied, the problem and solution for the resource allocation between the cells are also examined.

## 3.1 Telecommunication Services and QoS

Powerline communications (PLC) systems, such as the systems specified in [Opera06] or [HomePa07], support multiple users with multiple types of services. The telecommunication services are classified into different classes, based on the requirements of the QoS parameters.

The diagram presenting the data management with QoS support in a base station is depicted in Fig. 3-1. This block diagram has been defined in the PLC system described in [Opera06]. The diagram contains four main blocks: connection admission control (CAC), traffic classifier, buffer management, and traffic scheduler. The tasks performed by the blocks are explained as follows:

- Connection admission control (CAC): This block decides whether to accept or reject a new service request to a user based on the available transmission resources, QoS requirements of the new and current services.
- Traffic classifier: based on the type of the incoming request, this block assigns corresponding priorities to different traffic flows.

- Buffer manager: this block assigns the available memory to each flow and user depending on the users' statuses and types of the provided service. The data arriving from the traffic classifier are buffered in corresponding queues.
- Traffic scheduler: here, the data is packed into a corresponding data frame for the physical layer before transmitting them to the powerline. This block along with the CAC block calculates the point in time and period for accessing the PLC channel for the data in each queue for each user. The calculation module for these access parameters works based on a pre-defined scheme.

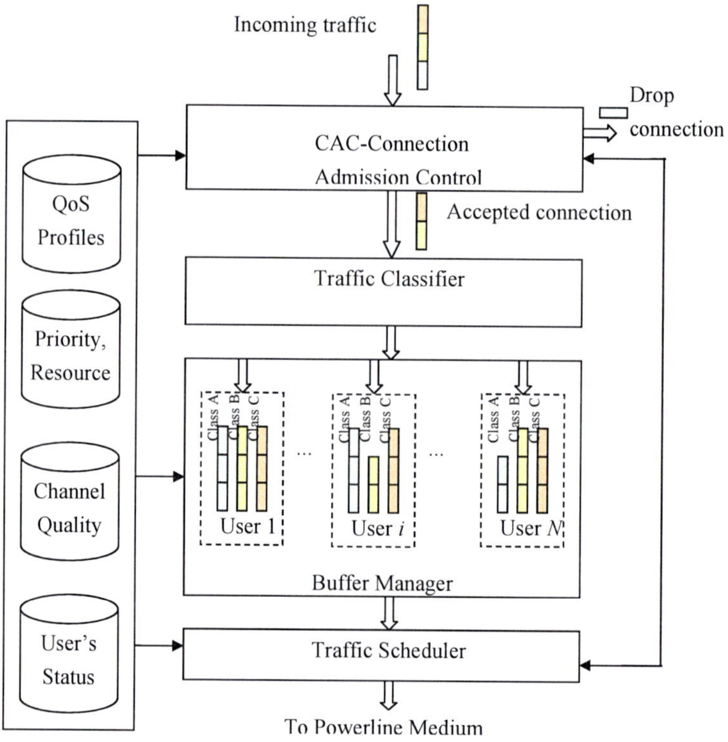


*Figure 3-1: Data management with different types of services [Opera06]*

For the time division multiple access scheme (TDMA) described in [Opera06], the transmission resources are the duration that a service is allowed to transmit. The decision for accessing the PLC channel is calculated based on the actual channel quality from the master to that user, the QoS parameter of the service, and other network parameters, e.g. other users' statuses and available transmission resources. Both the scheduler and buffer manager regulate the transmission resources and memory for delivering the services dynamically based on the current network traffic conditions.

The functions of the connection admission control module and the calculations for access time of each service are the main interests of the traffic scheduling. Generally, the mechanism should effectively use the limited available transmission resources to meet the QoS requirements from different services and different users. In this chapter, the priority assignment mechanisms for multiple flows are reviewed. It is assumed that data may only be discarded due to the congestion control mechanism, e.g. because data stays so long in a queue. The tasks for assigning resources to each service for each user are generally formed by the so-called traffic scheduling mechanism.

### 3.1.1 Telecommunication Services

#### 3.1.1.1 Service Classifications

The telecommunication services are classified based on service requirements. The requirements depend on the types of services, e.g. the traffic parameters' requirements for the data transmission are different from the requirement for video or voice transmissions [PaceAl05]. The number of types of services a system can provide depends strongly on the realization of the applied communication technique.

In accordance with [Opera06], four types of traffic are supported by a PLC system. They are constant bitrate traffic (CBR), variable bitrate traffic (VBR), available bitrate traffic (ABR), and best effort traffic (BE). Transmission resources for each service are reserved based on these kinds of the traffic flow. The requirements of each type of traffic are described as follows:

- The CBR traffic requires a constant transmission rate in a connection between two end nodes; e.g. in case of the master-slave PLC network, the connection is organized between the master and a user. Once accepted, this kind of service is always given enough transmission resources to constantly carry data.
- For the VBR traffic, the average bandwidth is guaranteed with a given maximum variation around a given average value. When this service is accepted, it is given enough transmission resources to carry requested data flows.
- For the ABR traffic, a given minimal bandwidth is guaranteed. A higher data rate may be provided over this minimum if there are free transmission resources.
- For the BE traffic, the traffic can be transmitted only when there is transmission resources remaining after all the other traffic has been scheduled. This kind of traffic is not given any guarantee.

Along with the above types of traffic, a PLC system in accordance with [Opera06] supports up to eight service classes. The service classes are defined by different network access intervals, service reliability, and traffic types. The service classes and parameters are presented in Tab. 3-1. In the traffic scheduler, a service with a higher service class has higher priority when different service classes are scheduled.

Since a PLC network is a shared medium system, all network equipment shares the same transmission medium to transmit the data. In a PLC system in which the time division multiple access (TDMA) scheme is used transmission resources of the users are defined by the base station as the time slots. The network access interval parameter defines the duration between two successive network access decisions. Different access intervals of the service classes are realized by using the time frame and sub time frame presented in Sec. 2.3. The transmission duration of a service is calculated based on the type of service and the current transmission rate of the PLC connection. The service reliability parameter describes whether acknowledgment is required for a service or not.

| Service class | Type of traffic | Network access interval (ms) | Service reliability | Example of the service |
|---|---|---|---|---|
| 7 | BE | 240 | No | Management messages |
| 6 | CBR | 30 | No | VoIP |
| 5 | VBR | 60 | Yes | Video, Games |
| 4 | VBR | 120 | Yes | E-commerce |
| 3 | VBR | 120 | No | Tele-services |
| 2 | ABR | 120 | Yes | Web traffic |
| 1 | BE | 240 | No | E-mail |
| 0 | BE | 240 | No | Low-priority Data |

*Table 3-1: Service class definition [Source: Opera06]*

#### 3.1.1.2 Service Quality Evaluation Methods

A service is evaluated by both the service provider and service user [ITUT08]. From the service provider's point of view, the quality of a service (QoS) is measured by means of the traffic parameters, in which the possibility of resource allocation for satisfying the service requirements is taken into account. On the other hand, the user, or subscriber, evaluates a service based on their opinion when experiencing the actual service; this is called quality of user's perception or quality of experience (QoE) [Laur96], [PiamVi09]. In this thesis, the main forcus is on the QoS parameters.

### 3.1.2 Priority Mapping for Multiple Service Flows

#### 3.1.2.1 Static Priority Organization and Mapping

When a request for a new service from a user is accepted, it is associated with a service class. The service class may be assigned a certain priority. In some telecommunication techniques, each service is mapped with a static priority [PaoLa98][DamoSi99]. This means that the priority of the traffic only depends on the type of service, not on the status of the transmission of that service in the network.

#### 3.1.2.2 Dynamic Priority Reassignment

In order to avoid a situation, in which a service in a lower class is blocked by the transmissions of services in a higher class, a change in the priority of the traffic stream based on the status of the current traffic is needed. This can be solved by several methods, e.g. one of them is described in [MoufNa06]: when traffic data queues longer than a pre-defined threshold, the priority of the traffic is increased.

## 3.2 Individual Performance versus Network Performance

### 3.2.1 Individual Performance Metrics

Performance metrics of communication services that a user receives from the network provider are data throughput, transmission delay, error ratio, service availability, and service reliability. A general analysis of these parameters is presented in a survey in [VogeKe95]. These parameters are also analyzed for different systems, e.g. for wireless system presented in [BordAn05]. In a system, these parameters are evaluated individually for each service. The performance parameters are either guaranteed or not guaranteed for a service provided to a user based on the type of service.

### 3.2.2 Trade-Off between Individual Performance and Network Performance

#### 3.2.2.1 Users' Performance Dependency

Since all the network equipment in a PLC network shares the transmission medium, the power lines, the performance of a user depends on the performance of the other users. Because the channel quality of the physical connection between a user and its master may be different from the other users, the sum of the transmission data rates in the network depends on the resource assignment to each user. Let us represent the channel quality on the PHY layer between a user ($U_i$) and its master is by the transmission data rate in that connection by $S_{U_i}^{PHY}$. We assume that the network has $N$ users and their transmissions are organized by means of the time division multiple access (TDMA) scheme. The durations of the medium accesses by each user and the corresponding transmission data rates with four users are presented in Fig. 3-2. When the master station reserves the same transmission capacity ($L$) for each users, the duration that each user requires to transmit is inversely proportional to the transmission data rates, because

$$S_{U_1}^{PHY} \cdot T_{U_1}^{A} = .. = S_{U_i}^{PHY} \cdot T_{U_i}^{A} = ... = S_{U_N}^{PHY} \cdot T_{U_N}^{A} = L \text{, and} \tag{3-1}$$

$$\sum_{i=1}^{N} T_{U_i}^{A} = T_{TF} \tag{3-2}$$

From (3-1) and (3-2), the dependency of the equivalent transmission data rate of the network in a time frame (number of transmitted data per time frame) and the transmission data rate of one user ($U_1$) can be given as

$$S_{Net} = \frac{N \cdot L}{T_{TF}} = \left( \sum_{i=1}^{N} \frac{1}{S_{U_i}^{PHY}} \right)^{-1} = \left( \frac{1}{S_{U_1}^{PHY}} + \sum_{i=2}^{N} \frac{1}{S_{U_i}^{PHY}} \right)^{-1} \tag{3-3}$$

Fig. 3-3 shows the dependency of $S_{Net}$ on $S_{U_1}^{PHY}$ based on (3-3) in a network with four users. This is assuming that the transmission data rate of $U_1$ varies from 1 to 40 data rate units while the other users ($U_2$, $U_3$, $U_4$) have a constant of 40 data rate units and the transmission capacity $L$ of all users are equal.

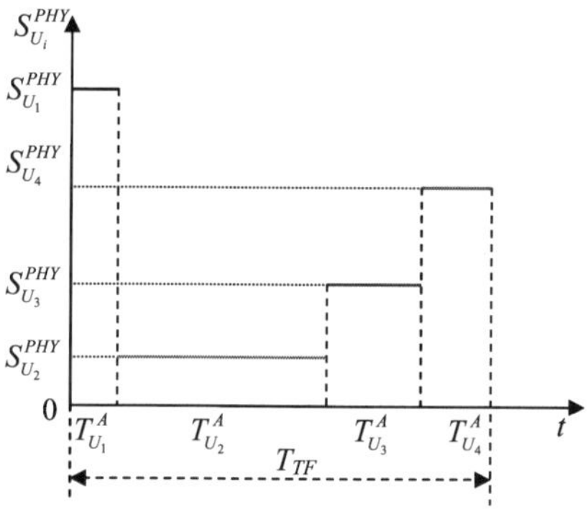


*Figure 3-2: Transmission durations of different users*

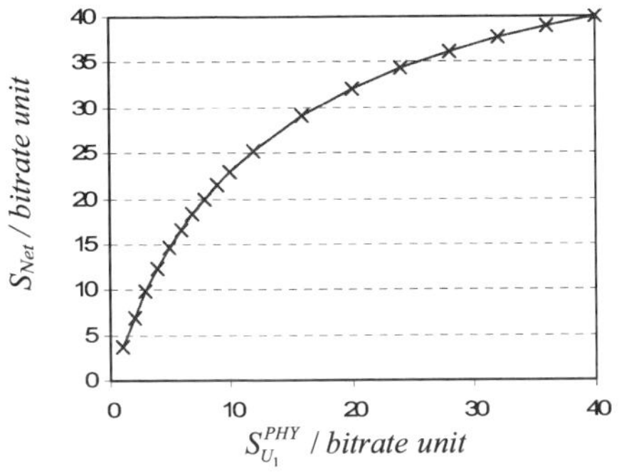


*Figure 3-3: Dependency of the equivalent network transmission throughput on the transmission data rate of one user (uniform transmission capacity)*

#### 3.2.2.2 Trade-off Principles

If the connection of one of the users has a low transmission data rate, guaranteeing the transmission capacity will require a high portion of the transmission resources. This will reduce the overall network performance. Let us consider a network with $N$ users as previously mentioned. If the network allocates enough transmission duration ($T_{U_i}^{A}$) for user $U_i$ for transmitting with the throughput $S_{U_i}^{Th}$, the portion of the transmission duration in a time frame is

$$\frac{T_{U_i}^{A}}{T_{TF}} = \frac{S_{U_i}^{Th}}{S_{U_i}^{PHY}} \tag{3-4}$$

When the channel quality between the user $U_1$ and its master changes, it is assumed that the user $U_1$ is allocated transmission resources to transmit with a constant throughput and the other $N$-1 users have the same transmission bitrates, the throughput of the other users are

$$S_{U_2}^{Th} = .. = S_{U_N}^{Th} = S_{U_2}^{PHY} \frac{T_{TF} - T_{U_1}^{A}}{(N-1)T_{TF}} \tag{3-5}$$

For the analysis, it is assumed that the network has four users and it allocates enough resources for user $U_1$ to guarantee the throughput at $S_{U_1}^{Th}$=10 data rate unit. Fig. 3-4 presents the transmission throughput of the other three users in accordance with (3-5) when the transmission data rate of user $U_1$ changes from 10 data rate units to 40 data rate units while the other are 40 data rate units. From the graph in Fig. 3-4, it can be seen that the throughput of the other users ($U_2$, $U_3$, $U_4$) strongly decreases when transmission data rate of the guaranteed user ($U_1$) decreases. The throughput of the users $U_2$, $U_3$, $U_4$ may drop to 0 when the user $U_1$ is guaranteed the same throughput as the transmission data rate. In this situation, the network assigns the entire time frame to user $U_1$.

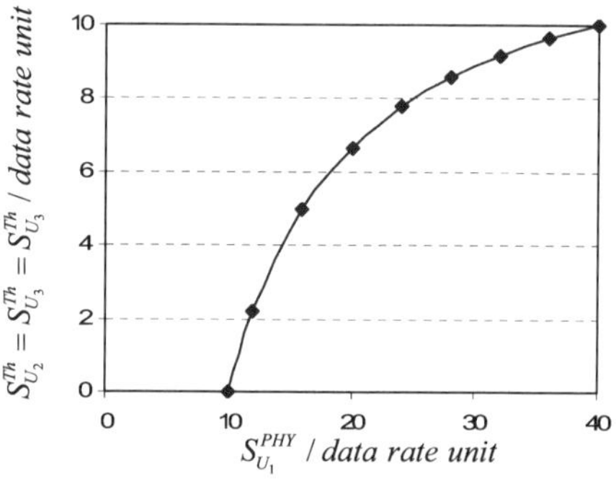


*Figure 3-4: Dependency of the transmission throughput of the other users on transmission data rate of user $U_1$ with throughput guarantee $S_{U_1}^{Th}$ =10 data rate unit*

Therefore, a trade-off between the individual performance of a user and the entire network has to be considered. This can be accomplished by

- limitation of the assignment of resources to each user. By doing so, the resource assignment to each user can be increased until it reaches the pre-defined threshold. The disadvantage of this method is that if the transmission data rate of a user is decreased too much, the services that a user utilizes may no longer be possible.
- dynamically adjusting the priorities of the transmission data of the users, e.g. by means of the proportional fair (PF) algorithm [Holt01]. With this method, the priority of the traffic from each user is proportional to the transmission data rate between that user and its master. If a user has a low transmission data rate, it traffic is scheduled with a low priority. Applying this method, a user with low transmission data rate is prohibited from consuming the resources for the other users that have higher transmission data rates.
- evaluation of the effective utilization of the transmission resources and fairness of the allocated transmission resources between the individual users in the entire network. The resource assignment task of the master station is formulated as an optimization problem. The problem can, therefore, be solved by optimization problem solvers. However, the optimization problem solvers may require a high performance calculation module.

## 3.3 Scheduling of Transmission Resources between Users in a Cell

### 3.3.1 Classifications

The scheduling of transmission resources between the users in a system is designed based on the characteristics of the communication system and requirements of the provided traffic services for that system. They can be classified based on wired and wireless systems [TsaiCh10]. In the medium access control (MAC) layer, there are two basic methods for multiple users to share a transmission medium in order to dynamically access the network: random access and dedicated access [Hras04]. As previously mentioned, the random access MAC protocols, e.g. ALOHA, carrier sense multiple access (CSMA), usually are used to provide low delay for short data. The dedicated access MAC protocol, e.g. token passing or polling, are normally utilized for networks with guaranteed data transmission. Dedicated access protocols can provide QoS guarantees for such services.

With the requirements from the mixture of various services with different QoS requirements over PLC networks, efficient schedulers are necessary. The scheduling schemes are evaluated by means of their most important parameters: efficiency, flexibility, and complexity [TsaiCh10]. The scheduling schemes allow the network to adapt to the demands of the users, maximizing the utilization of the transmission resources while taking into account the various QoS requirements of different services and maintaining a certain fairness between users and the services.

Based on the functionalities of the schedulers, they can be classified as presented in Fig. 3-5. This section reviews traffic schedulers that can be used in a master-slave system, which is an assumed principle of the referenced PLC network.

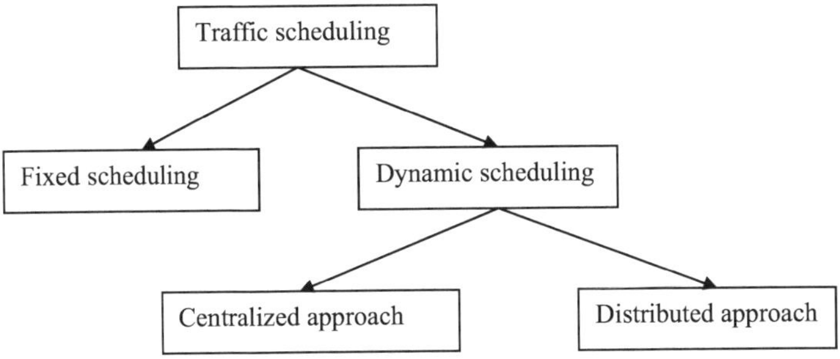


*Figure 3-5: Scheduling classifications*

#### 3.3.1.1 Static Scheduling

Static access in the MAC layer means that fixed transmission resources are assigned to each user. Examples can be the scheduling in slotted systems, in the frequency division multiplexing (FDM) networks, or wavelength division multiplexing (WDM) in optical networks [IshiMi84]. As depicted in Fig. 3-6, a base station is connected to three users via the same physical medium. However, each user is assigned a fixed frequency band to transmit data with its base station. The frequency band assignment is not changed during runtime.

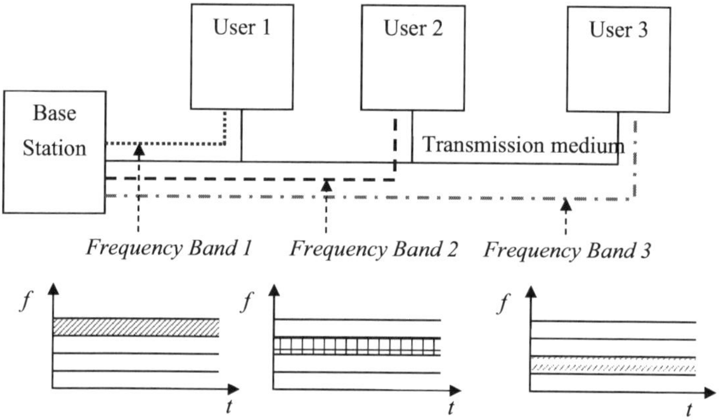


*Figure 3-6: Static scheduler with fixed frequency band assignments*

#### 3.3.1.2 Dynamic Scheduling

During the network run-time, the traffic requirement from each user may vary. Therefore, a scheduler that dynamically adjusts the allocated resources of each user is helpful. With the dynamic scheduling approach, the traffic resources are temporary assigned to a user in accordance with the requirements of that user and the availability of the traffic resources in the network. When the user no longer needs the traffic resources, these traffic resources are free to be assigned to the other users. An example of the dynamic scheduling can be found in several network technologies, in which a user is provided multiple services. One example is that the MAC layer utilizes time division multiple access (TDMA) with dynamic time slots assignment or a hybrid of FDMA and TDMA. Fig. 3-7 depicts the time slot assessments of the users in a network with a dynamic scheduler.

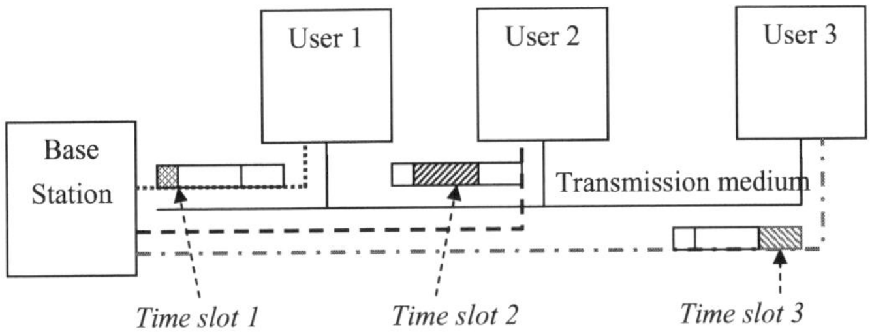


*Figure 3-7: Variable time slots assignment with dynamic scheduler*

The scheduling scheme can also be organized in a centralized or in a distributed manner.

#### 3.3.1.3 Centralized Scheduling

When the network is configured as a complete master-slave system, the master organizes transmission resources to each service in each slave. One advantage is that, only one unit takes all decisions – a unit, in which all the information about traffic requests and resource availability is known. The utilization of the available resources can be optimized with the traffic flow information. The disadvantage of a centralized scheduler is that this method requires a high performance calculation unit. This is because the scheduler has to perform complex calculations for the entire network, especially when serving for a network with huge number of traffic requests, large transmission resources, and within a given scheduling period.

#### 3.3.1.4 Distributed Scheduling

The distributed scheduling, on the other hand, has not only the main scheduler in the base station but also schedulers in the slave stations, e.g. as the principles of the schedulers in both uplink and downlink traffics in WiMAX [WiMAX04] or the schedulers for passive optical network [MastPa05]. Generally, the scheduling contains two parts: a master scheduler and slave schedulers. Both the master scheduler and slave schedulers actively participate in making the resource decision for a new service request. For a master slave system, each user has one slave scheduler. The master scheduler allocates resources to each user in accordance with their total traffic request instead of individual service requests. When a certain amount of transmission resources is allocated to a user, the slave scheduler in that user schedules the internal services. By fragmenting the scheduling tasks into two sides, the complexity of the calculation in the master is reduced because of the participation of the calculation units by all users. However, the disadvantage is that the resources for a specific traffic may not be guaranteed from the scheduler in the master. Fig. 3-8 presents flowcharts of message sequences between the base station and a user before starting a service in both centralized and distributed schedulers.

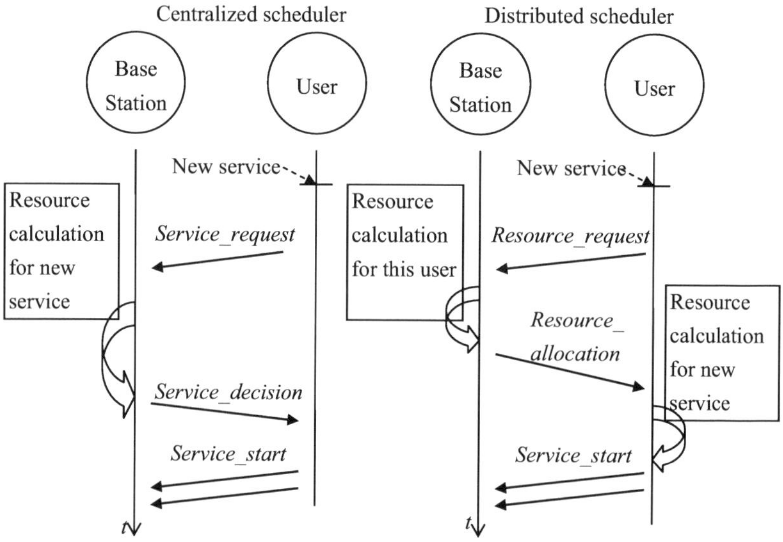


*Figure 3-8: Flow-charts of centralized and distributed schedulers*

### 3.3.2 Traffic Control

Besides the task of scheduling the services in the network, the traffic resource manager in the master station has to control the transmissions during network operation. This consists of the congestion control, error control, and queue monitoring and drop control.

#### 3.3.2.1 Congestion Control

There are several proposals and mechanisms for controlling the congestion over a link in the network. This can be realized in different communication layers [LuoLi01][WidmDe01]. As introduced in [Opera06] for the PLC networks, congestion control is necessary in case there are changes in the network, e.g. link quality changes between two successive scheduling procedures. In the MAC layer, solutions for controlling in the case of network congestion are:

- fairly reducing the qualities of all the current services or
- strongly reducing the quality of elastic traffics while maintaining resources for real time services.

With a fair congestion control policy, allocated transmission resources for all of the services are reduced proportional to the allocated resource. This method may cause the real time services to no longer be useable, e.g. VoIP with transmission resources lower than requested resources for a certain threshold of throughput. On the other hand, the methods of priority congestion control takes the priorities of the services into account while adapting the resources to the changes of the channel qualities by only reducing the resources for services with lower priorities, e.g. elastic services. This method is acceptable for a user with a combination of both elastic and real time traffic. However, it may produce unfairness between users when there are users with only elastic traffic.

#### 3.3.2.2 Error Control

For dealing with errors of the transmission medium, mechanisms, e.g. forward error correction (FEC) and automatic repeat request (ARQ) may be applied. Utilizing the FEC technique, the transmitter encodes data with redundancy. Using this redundancy, the receiver, can detect and correct a certain number of errors that may occur in the received message. The ARQ utilizes acknowledgment. In some systems, this acknowledgment method can be realized by means of a dedicated acknowledgment message or a piggy-backing method, where acknowledgment is sent from the receiver to the transmitter along with the data in reverse transmission direction [Hras04]. With the information from the receiver via acknowledgment, the transmitter can retransmit the lost data. The acknowledgement and retransmission of lost data may be activated based on types of services and requirements. For some types of services, e.g. real-time services like video streaming, retransmission of the missing packet is not required, as presented in Tab. 3-1.

#### 3.3.2.3 Queue Monitoring and Drop Control

Queue monitoring and management are necessary for controlling the traffic, especially when the qualities of the connections dynamically vary during network operation. In the master station of a master-slave system, the buffer manager module distributes the buffer memory available among network nodes and the services. In a node, the queue manager controls the sharing of the buffer between the services. The queue manager module also regulates the queue length of a service based on characteristics, e.g. throughput, and delay limit of that service. The data packet, therefore, may be dropped when its corresponding queue is full. In this investigation, it is assumed that the packet dropping occurs simply because of the service requirements. It makes no sense to transmit data packets of real-time services which have been waiting in the queue longer than the maximum delay, e.g. if a VoIP data packet arrives too late at the receiver - at the time when it should have left the play-out buffer.

## 3.4 QoS Aware Scheduling

Considering the communication in a PLC network with centralized resource control, the scheduler has to organize the traffic from and to different users within the limited available transmission resources. Each user may also request different amount of resources for a different number of requested services. If the sum of the resource requests is lower than the available resources, the scheduler can allocate enough resources for all requests. This can simply be realized by the well-known first come first serve (FCFS) scheme in which the earliest arriving data is served first. However, the FCFS scheme is not suitable when there are different types of services with different priorities. Furthermore, when the users request more resources than available, a strategy is required to allocate the limited resources.

Different scheduling mechanisms have been used for different communication systems [TsaiCh10], [Alfy10]. The following section reviews scheduling schemes that support QoS.

- Priority queuing (PQ): data from each service is buffered in a specific queue and each queue is served according to its assigned priority. Data are processed from the queue starting with the highest priority to the lowest priority. When there are several services with the same priority, they are served, however, based on the first come first serve (FCFS) principle. This is known as HOL (head-of-the line) priority scheme. If there are always high priority data to be transmitted, no resources will be remaining for the data in the queue with lower priority. This situation causes unfairness between the services.
- Round robin (RR) and its variants: With the round robin scheme, the available transmission resources are uniformly granted to each user in a cyclic order. A variant called weighted round robin scheme (WRR), additionally assigns a weight to each queue in accordance with the priority of the service. The amount of data to be served from a queue is proportional to the weight of the queue, thus this method can provide a certain fairness between different services. Other variants of the round robin are the deficit weighted round robin schemes (DWRR) [ShreVa96][GregSh00]. In this scheme, the number of bytes to be processed from each queue is controlled by a deficit counter. This counter allows the scheduler not only to control the priority of each queue but also the number of bytes to be served under different network conditions.
- Earliest Deadline First (EDF): As previously mentioned, for real-time services, the deadline for delivering a packet may be hardbound. This type of scheduling has been used for several systems, e.g. to provide different multimedia services to users [AntoSt97] and different real-time streams [HoanJo02]. With the EDF scheduling scheme, the data which has the nearest deadline will be served first. The disadvantage is that other services, e.g. best effort traffic, may be blocked when there is always real-time data near deadline to be transmitted.
- Proportional Fair (PF) algorithm: In a network where the data rate in a connection between a user and its master is different from other connections, the allocation of the same transmission resources to all users results in different transmission capacities

between users. With the principle previously mentioned in section 3.2.2.2, the proportional fair algorithms gives a trade-off between the entire system performance and fairness of the users by granting the traffic resources to the user proportional to its transmission data rate. The fairness between the users is better than using the method of granting the same transmission capacity to the users while the system throughput is still high [SorePo05].

- Opportunistic scheduling: Generally, the opportunistic scheduling uses the channel quality in the process to determine the resource allocation. The idea of this scheme is presented in [TseHa98] and further investigated, e.g. for scheduling in an OFDM downlink [ZhanHe05]. However, in accordance with the analysis in Sec. 3.2.2, if the scheduler grants too many resources for a user with a very low transmission data rate, the total system throughput will be low. Depending on the objective of the allocation scheme, the scheduler changes the appropriate traffic resources for each service of each user. Several systems, e.g. as proposed in [ZhanHe05][LeiLi09], use this method for optimizing the utilization of the limited available transmission resources.

## 3.5 Resource Sharing between Multiple Service Providers in a Single Cell

### 3.5.1 The Model

Even physically located inside a cell, users may be associated to more than one service provider (SP). In a PLC network, a telecommunication service provider and a power utility may share the same PLC network for providing communication services and electricity management to the users. For a PLC network, these scenarios have been analyzed in [D13OP07] with the so-called carriers' carrier model. The carriers' carrier model is applied when several communication service providers buy or rent bandwidth from an infrastructure provider. These service providers have to share the same infrastructure and provide their services to their own customers. Each customer may freely choose his service provider.

Following the model in [D13OP07], the PLC is built by the electrical power supplier, which provides a virtual network to each service provider (SP). All the users of an SP are assigned to a so-called virtual private network (VPN). The electrical power supplier that provides the VPN access is called a VPN provider. The VPN provider designs the network infrastructure, provides backbone access, and maintains the total network infrastructure. The service provider provides its users value added services, e.g. VoIP, data transfer, to its users. Each SP can provide specific offers to its users, e.g. different maximum transmission data rates at different subscription prices.

An example of a PLC network with two VPNs is shown in Fig. 3-9. A base station (BS) is installed in the transformer station to provide network access to the users located inside the low voltage network. The BS connects to the user directly or via repeaters (RP) and connects itself to the backbone network (backhaul).

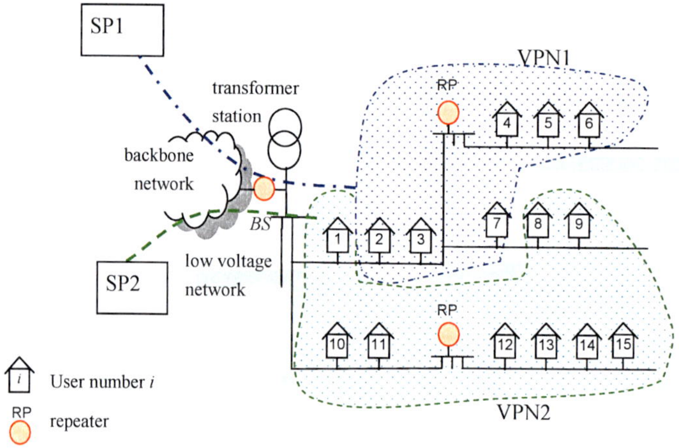


*Figure 3-9: PLC network with multiple VPNs*

There are different methods for configuring multiple VPNs for multiple service providers as specified in [D13OP07][QinLi02][HondOh05], including IP (Internet Protocol) range selection, virtual local area network (VLAN) tagging, or grouping of traffics from users belonging to an SP via an access port. Because the PLC network is a master-slave system, the BS aggregates all traffic from its slaves and sends to the corresponding SP in accordance with the configuration. All users associated with a VPN will be served by the SP via the management in the base station. Using this model, the sharing mechanisms for the transmission resources between the service providers are studied.

### 3.5.2 Traffic Sharing between Service Providers

When the available transmission resources are not enough to satisfy all the requirements from the users in one or more service providers (SPs), a sharing strategy has to be applied. When applying the carriers' carrier model, the VPN provider has to maintain the scalability of the bandwidth between SPs. Each SP has a specific "weight" reflecting its priority when sharing resources because of e.g. its local topology or contract with the VPN provider. The incoming traffic in the downlink direction, after being accepted by the connection admission control, will be put into different queues based on the destination and type of service by the traffic classifier block and for each SP. The traffic scheduler serves the queues based on the class of service parameters and the weights of the SPs.

Among several sharing strategies presented in [JonaGu02] and [Welz05], one well-known strategy that can be used by the BS in this scenario is the weighted round robin method. On the other hand, the scheduling for multiple VPNs in which the resource allocation problem is formulated into an optimization problem is proposed in [DoLe09]. The principle of identifying the data packets from and to different SPs is described in [D13OP07]. The MAC data packet to be transmitted over the power line is equipped with a field to carry the VLAN-Tag. In the downlink direction, from BS to user, when an Ethernet packet is translated to the corresponding packet in the MAC layer, the access port of the SP is translated into a corresponding VLAN-Tag. Only users who are assigned to the corresponding VPN can receive the packet with the correct VLAN-Tag. In the uplink direction, from a user to the BS, the MAC data packet is translated back to the Ethernet packet and the VLAN-Tag is translated to the corresponding port of the SP.

## 3.6 Transmission Resources Sharing Between Multiple Cells

To cover a large area, a network may be configured into multiple smaller cells. This cellular principle known from mobile networks is also investigated for PLC systems. Considering the interference conditions between the adjacent cells in a network, a method to allocate the limited transmission resources into each cell is required. The transmission resources can be fragmented into channels assigned to the cells. As known from mobile networks, e.g. GSM systems, the channel assignment techniques may be classified into two categories, fixed and dynamic allocation methods. Fig. 3-10 presents the classifications of the channels allocation methods. Some studies have been conducted and solutions proposed for this problem in the PLC networks as well, e.g. in [Haid07] and [GallKu09]. This section analyzes the available channel allocation solutions which can be applied to multiple cell PLC networks.

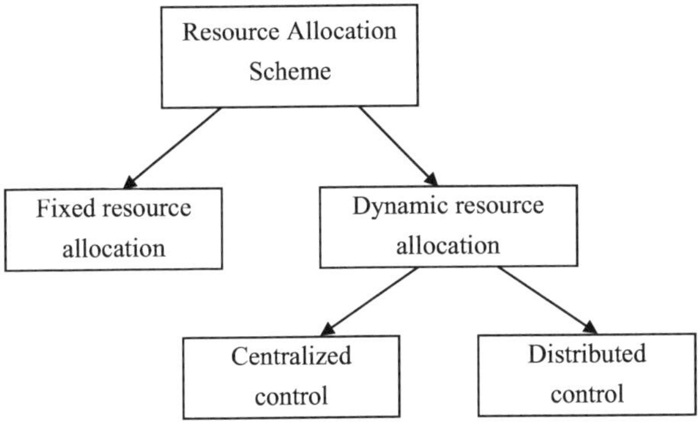


*Figure 3-10: Classification of resource allocation scheme for cellular network.*

### 3.6.1 Fixed Channel Allocation between PLC Cells

This method requires full knowledge of the network topology and the traffic demands from the cells in the entire network. In this method, the channels are assigned to individual cells with the goal to operate over a long period. Since complete knowledge is required, the allocation is normally performed when the network is planned. The allocation can also be done by optimization algorithms [ChenSe01] [Wong03]. There are investigations that have proposed to share the transmission resources between the cells in PLC network in a statistical fashion; e.g. [Haid07] has proposed to solve the channel allocation problem when planning the network.

One example of the solution for the FCA is shown in Fig. 3-11. The network depicted consists of four cells ($\boldsymbol{C}=\{C_i\}/i=1..4$) and assumes that five channels ($\boldsymbol{F}=\{F_j\}/j=1..5$) are available. As presented in Fig. 3-11, transmissions in cell $C_1$ and cell $C_2$ may interfere with each other; therefore they are not assigned the same channels. Cell $C_3$ is located far from cell $C_1$, hence it is assigned again channel $F_1$ as well as which is used in $C_1$. This allocation of channels in each cell may be optimal in the time of network planning, but the fixed number of allocated channels leads to unfairness between the users in different cells when the traffic demand from users varies.

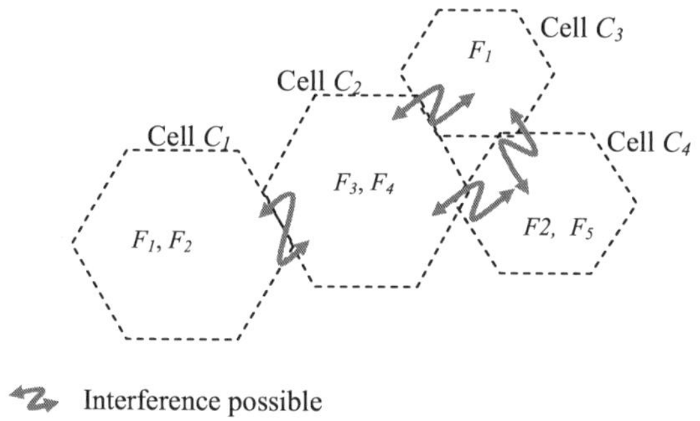


*Figure 3-11: PLC cells with fixed channel allocation scheme*

The advantage of this approach is that several calculation methods can be applied including exhaustive search or heuristic algorithms because the allocation algorithm does not have to run in real time. Therefore, the allocation for the planned traffic may be optimal. However, the traffic demand, as input for the problem solver, is a forecast from each cell and may differ from the actual traffic demand during run-time. Thus, the applied channel allocation solution is no longer optimal during run-time.

### 3.6.2 Dynamic Channel Allocation between PLC Cells

Dynamic Channel Allocation (DCA) is a method in which the channels are not permanently assigned to a certain cell, but are allocated to the cells according to the dynamic traffic demand from them. The channels are resource units that can be moved from one cell to another during run-time. There are several studies on the algorithms for dynamically solving the channel allocation problem, but most of them were originally proposed for wireless cellular system, e.g. for GSM in [AryrJo99] or [AnanSr03]. The dynamic channel allocation (DCA) can be implemented centralized or distributed. In the centralized approach, the requests for channels have to be transmitted from all cells to a common central controller. The controller estimates the overall network traffic demand, assigns channels to the cells, and informs the cells about the allocated channels. Using a distributed approach, all of the cells take part in the decision to determine the channel allocation in the network. For the master-slave model of a cell, each base station (BS) has to make decisions based on its own knowledge and the information from its neighbor cells.

#### 3.6.2.1 Review of Methods for Solving the DCA Problems

Several algorithms have been introduced for solving the DCA problem. For cellular GSM networks, a number of studies for the DCA have been conducted, e.g. as reviewed in [FranNi97] and [LoiWi99]. In addition, the particular characteristics of the GSM network have been taken into account; for example, [SamlGi01] proposes to dynamically distribute frequency bands between the cells by the so-called DFCA (Dynamic Frequency and Channel Assignment), while [WenCh01] and [LozaDo02] propose algorithms for solving DCA between the cells by means of the TDMA scheme for mobile communications systems.

Among some well-known DCA algorithms like the autonomous reuse partitioning scheme [Kana92] and minimum interference [ShinCh99], various solutions have also been proposed using other characteristics of the GSM systems. In [AndFr95], the authors use the geometric information in the strategy while solving the DCA with the bursty user mobility. The idea proposed in [BoukHo00] and [LiniDo05] is to solve the DCA with a distributed approach. The authors in [BubMa07] define and evaluate the DCA solutions by a cost function based on the provided services, the traffic load, and interference between the cells.

In other areas of telecommunications, some studies to solve DCA problems have also been conducted. The proposed solutions are based on the characteristics of the communication technologies, e.g. the solutions have been proposed by [MastPa05] for Ethernet Passive Optical Networks (EPONs).

Game theory has also been applied to solve the DCA problem. The methods of game theory and their applications in the communication areas have been reviewed and analyzed in [FeleHu07]. It has been used, e.g. in [ZhonXu05] for power allocation in MIMO (Multiple

Input Multiple Output) channels and for power control in CDMA (Code Division Multiple Access) systems in [MeshCh05]. Generally, each game consists of several players that try to achieve certain objectives, called payoff [Stin02]. In a problem which is defined as a game, the solution is conducted when the players decide to perform certain actions at the same time or consecutively. The payoff of the solution depends on the combination of the actions that the players performed. The authors in [NiyaHo07] use dynamic game theory and in [SuriDa07] use cooperative games for spectrum sharing. [WongWa02] applies game theory to solve the DCA problem for the multiple cells wireless networks.

#### 3.6.2.2 Centralized Approach

For the centralized approach, a central controller and its connections to the base stations in all cells are necessary, as depicted in Fig. 3-12. This form of organization for the central controller is possible if the network is owned by one or a cooperating network service provider.

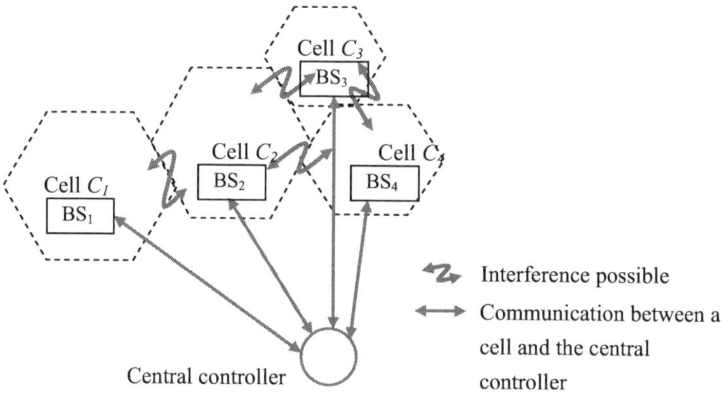


*Figure 3-12: Centralized dynamic channel allocation*

The information is periodically exchanged between the cells and the central controller. All the channel information, e.g. channel status, capacity of each channel, is stored in a database of the central controller. The network structure, e.g. described by the interference between the cells is also stored in this unit. The cells regularly send their channel requests to the central controller. Based on the available channel information, the controller estimates the overall network traffic demand, calculating channels in each cell, and informs the cells of the new allocated channels. A message sequence chart (MSC) demonstrating the process of the centralized dynamic channel allocation in the central controller is presented in Fig. 3-13.

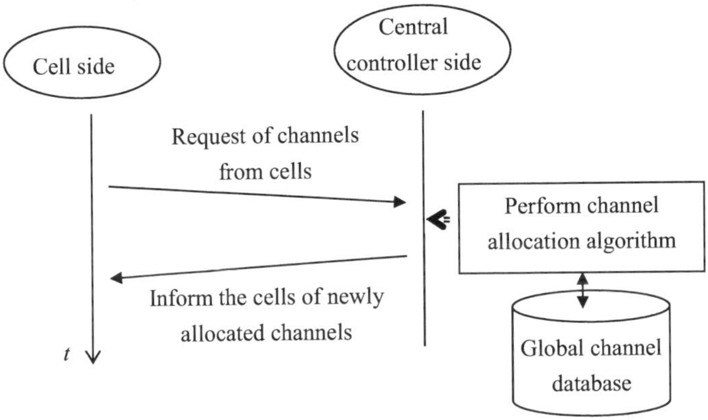


*Figure 3-13: MSC of the process for centralized dynamic channel allocation with the central controller*

This method requires a relatively fast decision and response from the central controller [PapzKa05].

#### 3.6.2.3 Distributed Approach

A distributed approach becomes necessary when the centralized dynamic channel allocation is not applicable, e.g. due to large-scale networks with too many cells. In the situation of a PLC network, the cells may have different owners or service providers, meaning that organizing a common central station for the entire PLC network may not be possible. In the distributed approached, each base station has to participate in the decision for the allocated channels of its cell and its neighbor cells. This has to be performed without a common central station. In this case, the base stations are not connected to a common unit but only to its neighbor cells to exchange the channel information and to make decisions on the channel allocation. A common algorithm with a set of rules has to be applied to each cell allowing them to collaborate.

Let us consider an example network consisting of four cells as presented in Fig. 3-14. The process for the channel allocation in this situation has to be made locally at each base station, as depicted in Fig. 3-15. Cell $C_2$ is a neighbor of the cells $C_1$ and $C_4$, therefore, its channel decision will effect to the channels used by the cells $C_1$ and $C_4$. If cell $C_2$ seizes channels $F_3$ and $F_4$, $C_1$ and $C_4$ should only use the remaining channels in order to avoid possible interference. However, the decisions of the cells $C_1$ and $C_4$ have to be made regarding the channels in use in the cell $C_3$, otherwise interference between cell $C_1$ or $C_4$ and $C_3$ may occur.

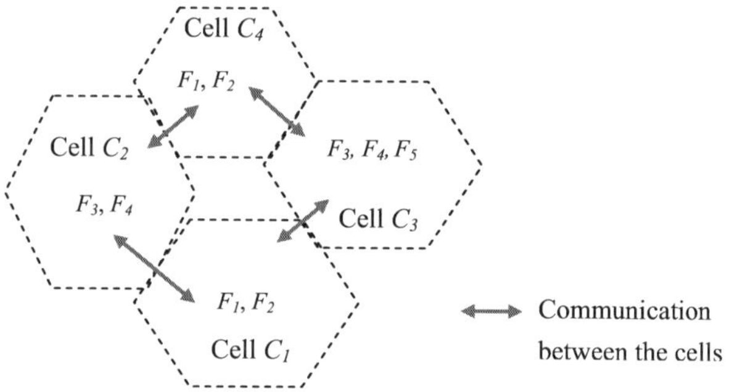


*Figure 3-14: Multiple cells with distributed channel allocation scheme*

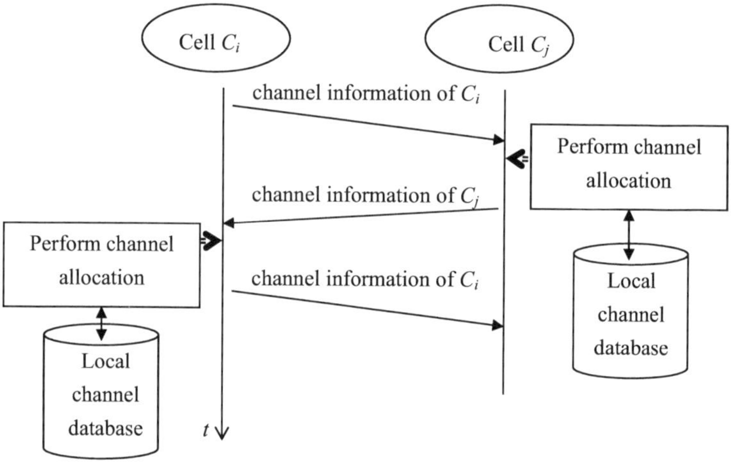


*Figure 3-15: MCS of the distributed dynamic channel allocation scheme*

# Chapter 4 PLC Network Model

The powerline communications (PLC) network is planned and installed in an area to provide telecommunication services to the users in accordance with the actual available infrastructure of the electrical power lines in that area. To cover a large area, the entire PLC network is divided into multiple PLC clusters, called PLC cells. Each PLC cell is a master-slave system, where one base station is installed to serve all connected subscribers. All the network devices share the same physical channel to access the network.

## 4.1 The PLC Network Structure

The PLC network is usually planned to efficiently serve all users within the required area. Algorithms are applied to specify the logical network, e.g. with the planning procedure presented in [HrasHa04] or in [Haid08]. For the planning process, the inputs are the current power line network infrastructure, possible positions for installing a base station, and positions and traffic requests from the users, as presented in Fig. 4-1a and Fig. 4-1b, which give an example of a PLC network constructed over the low voltage grid as the result from a planning task. The entire network consists of several base stations installed in some selected street cabinets, zero or several repeaters, and the users. Users in the network are assigned to a corresponding base station to form a cell. For efficient utilization of the powerline, users are usually assigned to a base station nearby. In case a user is too far from the base station, a repeater may be installed between them. The example logical network in Fig. 4-1b consists of 5 cells.

During the network operation, the physical powerline network may be changed; therefore, adjustment to the logical network structure may be necessary. The network can be re-arranged by means of a re-planning process, or by means of a network self-organization procedure. Network re-planning or self-organization processes are assumed to be completed by another special procedure, e.g. utilizing the automatic master selection protocol in the OPERA specification [Opera06]. The logical network structure, available transmission resource, and traffic requirements are used as inputs for the resource allocation process. A certain number of transmission resource units are required by each cell to meet the traffic requirements inside that cell. In case the number of allocated resource units is lower than the required amount of resources for the given traffic, the resource allocation protocols have to manage the allocated resources to ensure high resource utilization by the users while preventing interference not only inside that cell but also with the other cells.

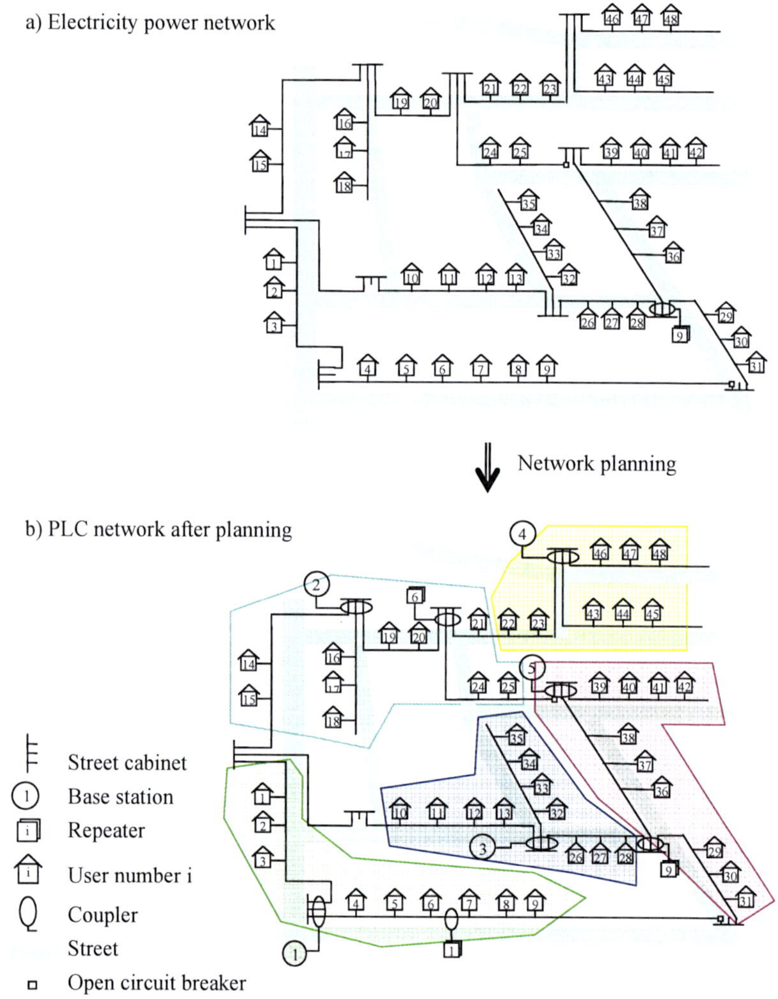


*Figure 4-1: PLC network in a low voltage grid after network planning*

## 4.2 PLC Logical Network Model

### 4.2.1 Reference Models for Multiple PLC Cell

As previously mentioned, the cellular concept is applied to PLC systems to provide services to a large number of users with limited transmission resources available. A multi-cell PLC network formed from several adjacent PLC cells may utilize different PLC technologies; one uses OPERA [Opera06] while another uses Homeplug [HomePl06] specifications. A sample cellular model of with multiple cells forming the so-called multi-cell PLC network is depicted in Fig. 4-2. Inside a cell, network devices are connected to each other via low voltage (LV) electricity power lines. Each cell is a complete master-slave system, consisting of one base station (BS), zero to several repeaters (RP), and users. The base station is the master in a cell and is itself connected to the backbone network via a high-speed connection, e.g. optical link or Very High Speed Digital Subscriber Line (VDSL), in order to provide access services for its users, directly or via repeaters. The cells that are able to interact with each other are combined to form a separate PLC network.

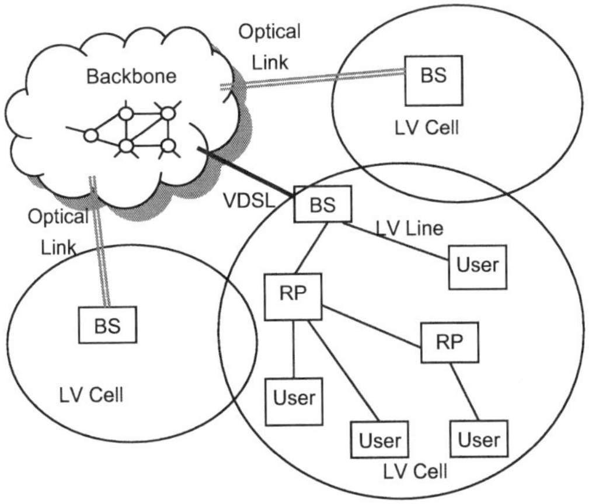


*Figure 4-2: Logical PLC network with a multi-cell structure*

Multi-cell PLC network topologies are based on the user distribution as well as the electrical power line infrastructure. Let us consider the distribution of users in a site where a service provider wishes to provide telecommunication services. The distribution of the electricity customers can be categorized to one of following typical scenarios [Sche02]:

- High density of customers in several wall-to-wall medium-height buildings. Each customer may locate in a building.
- High density of customers in several short-distances medium-height buildings
- High density of customers in a high-rise building. Several customers may locate in a floor.
- High density of customers in small houses

- Medium density of customers in short-distance high buildings
- Small group of customers in a village
- Low density of customers in high-distance small houses

Throughout these distributions of customers in the power line networks, three logical PLC multi-cell structures are considered to be the reference networks. They are derived from European power grid topologies proposed in [D13OP07] and [D16OP05], and are briefly presented in the following sections.

#### 4.2.1.1 Multiple PLC Cells in a Chain Structure

As presented in [D13OP07] and shown in Fig. 4-3, this network structure is formed by the neighbor cells located in a chain. Typically, this kind of network is constructed over long power lines with the users in small houses located on a long street. This kind of network is modeled by a chain of the PLC cells, as presented in Fig. 4-4.

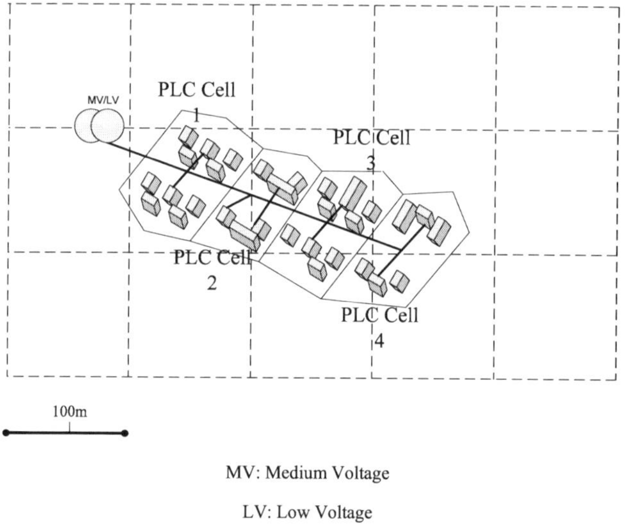


*Figure 4-3: Example of a group of customers in long street [Source: D13OP07]*

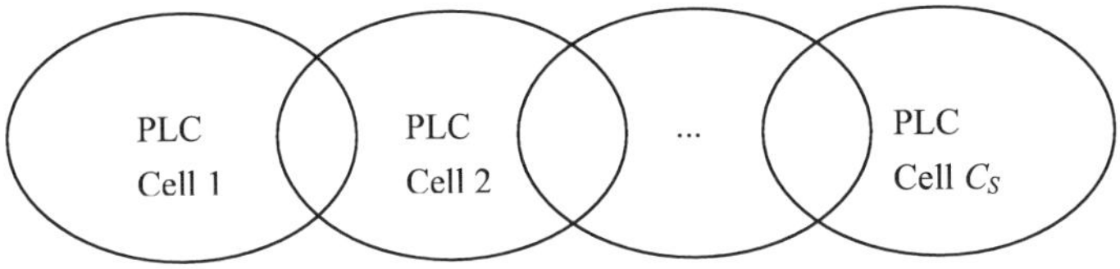


*Figure 4-4: Logical PLC network in a chain of cells*

#### 4.2.1.2 Multiple PLC Cells in a Ring Structure

In a ring-type organization of the LV networks, the powerline cells built over the LV network can also be organized in a ring structure. One example of this network structure is the organization of the high-density users in the small houses [D13OP07] shown in Fig. 4-5. A model of the network structure with *S* cells located in a ring topology is presented in Fig. 4-6.

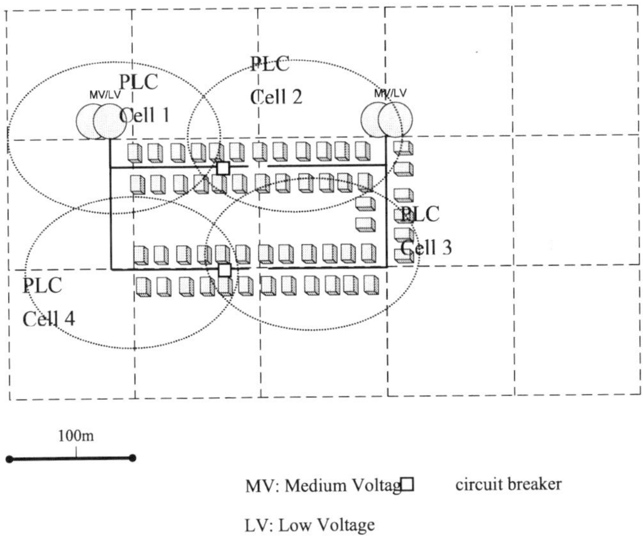


*Figure 4-5: Example of a high density of customers feed by ring power lines [Source: D13OP07]*

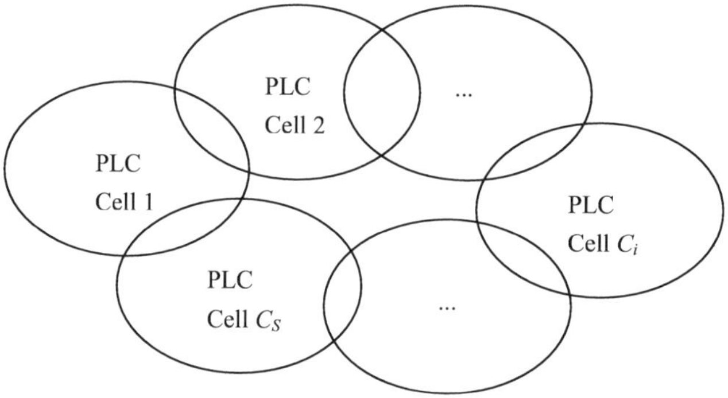


*Figure 4-6: Logical PLC network with ring of cells structure*

#### 4.2.1.3 Multiple PLC Cells in a Meshed Structure

All the other configurations of the PLC network are categorized into the so-called meshed network structure. A sample network with five PLC cells, taken from [D13OP07], is presented in Fig. 4-7. A meshed network can contain of several parts, each can be in a chain, a ring, or a star structure. A corresponding logical network structure of the meshed multi-cell PLC network is presented in Fig. 4-8.

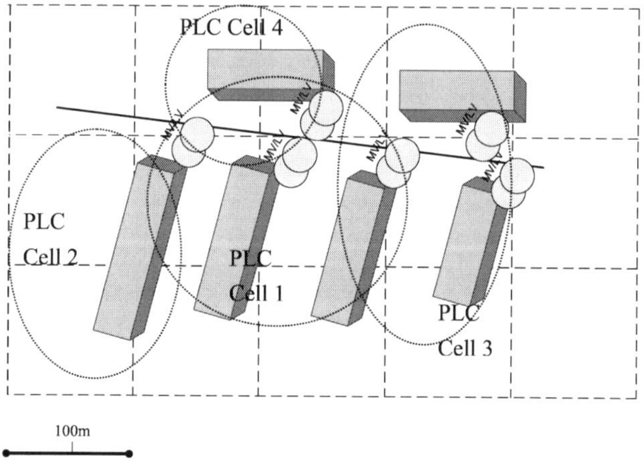


*Figure 4-7: High density of customers in several short-distance medium buildings [Source: D13OP07]*

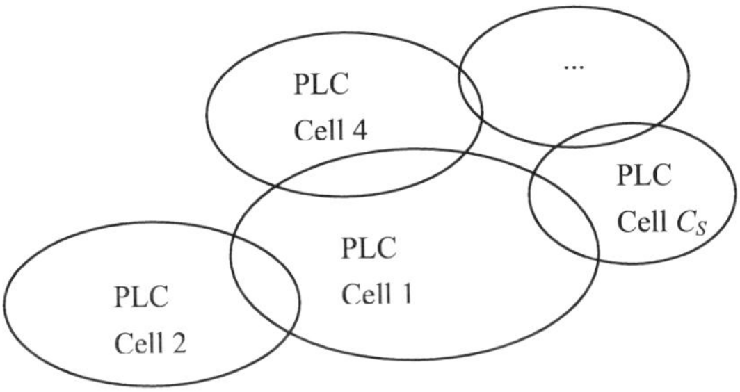


*Figure 4-8: Logical PLC network in a meshed structure*

### 4.2.2 Reference Models for the Single PLC Cell

Three PLC cell models with different network sizes, referred to as small size cell, medium size cell, and large size cell, are considered to be used for evaluating the resource allocation procedure inside a cell. They are described in the following section.

#### 4.2.2.1 Small-Sized PLC Cell

A small cell contains a base station and only 8 users, located in a small area. Via short distance power lines, all users are connected directly to the base station.

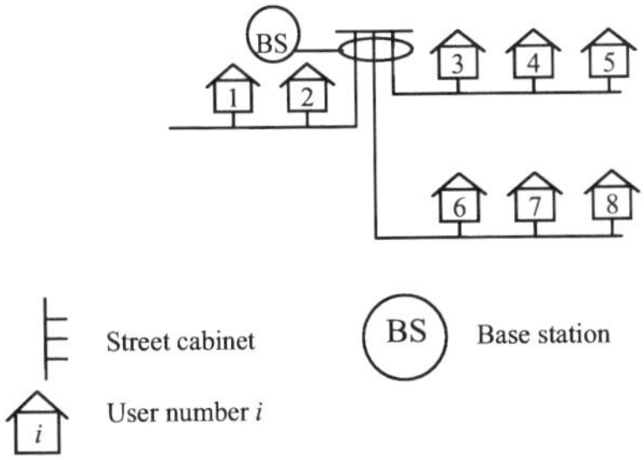


*Figure 4-9: A small PLC cell with users located around a base station (BS)*

#### 4.2.2.2 Medium-Sized Cell

For further investigation of the performance of the resource allocation scheme, a medium-sized PLC cell which has 25 users connected to a base station via 4 repeaters is taken into account. Channel qualities of the connections in the network and the service requests from the users are presented in chapter 7.

#### 4.2.2.3 Large-Sized Cell

A larger PLC cell which contains a base station, 120 users, and 7 repeaters can also be used to investigate the performance of the proposed resource allocation protocols. The network topology is generated based on the topologies of an electrical networks presented in [D11OP07].

## 4.3 Assumptions for the Studied PLC Technologies

The resource allocation scheme is proposed to work with different PHY and MAC layers. This is also a requirement for the co-existence between different PLC systems that utilize different PLC technologies. In the reference model, it is assumed that the PHY and MAC layers are existing and give access to the network based on the output from the scheduler running the proposed resource allocation scheme. Because there are particular specifications for PHY and MAC required for the PLC technique, the OPERA specifications [Opera06] are used for our evaluation of the proposed resource allocation method. Generic requirements for the other specifications are also presented and analyzed.

## 4.4 Logical Connections between PLC Cells and User Model

### 4.4.1 Logical Connections between PLC Cells

In order to investigate the resource sharing protocol between the cells, a representation of the logical connections between the cells is necessary. It is assumed that the most robust modulation scheme is used for the transmission between the cells and the quality of a logical connection is represented by the packet error ratio (PER). PER is defined as the number of erroneous packets normalized by the number of sent packets. This parameter depends on the characteristics of the transmission channel and the data packet length. The data is sent from the BS of the transmitter cell to the BS of the receiver cell. Therefore, connection between cell $C_i$ and cell $C_j$ implies the end-to-end connection between two BSs of two cells.

Since each connection between two cells is represented by a PER value, the entire set of logical connections in the investigated network therefore can be represented by a PER matrix. Each element PER$(i,j)$ represents the packet error ratio between the cells $C_i$ and $C_j$. By utilizing the OFDM (Orthogonal Frequency Division Multiplexing) modulation scheme, the packet length can be represented by a number of OFDM symbols. In accordance with [D51RE03], each PER of a connection can be calculated based on the signal to noise ratio (SNR) and the number of bits per symbol ($N_{sym}$) for the transmitted packet over that connection. For an evaluation of the information exchange between the cells, it is assumed that the PER matrix of the network is already known.

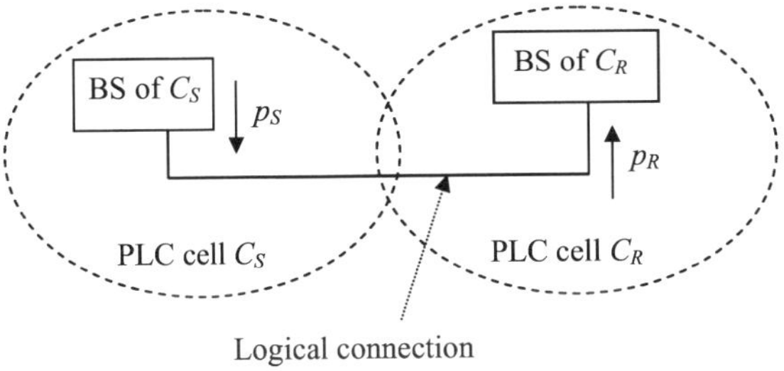


*Figure 4-10: Connection between two PLC cells*

If we observe a transmission between a transmitter cell $C_S$ and a receiver cell $C_R$, as depicted in Fig. 4-10, the probability of receiving a transmitted packet at $C_R$ in a time slot is given by the following equation:

$$p_R = p_S \cdot \left(1 - \mathrm{PER}\left(C_S, C_R\right)\right) \tag{4-1}$$

where $p_S$ is the probability that a transmitted packet is sent from transmitter cell $C_S$ in a time slot and PER($C_S$, $C_R$) is packet error ratio of the connection between two cells, $C_S$ and $C_R$.

This formula is correct only there are no parallel transmissions to the cell $C_R$. When there is a cell $C_{S^-}$ nearby, which also has a logical connection to the cell $C_R$ and sends its packet with $p_{S^-}$ in a time slot, it produces interference to the reception of $C_R$ when it sends its packet in the same time slot as $C_S$. In general, cell $C_R$ can only receive the packet from $C_S$ correctly if no other packet has been sent from interfering cells:

$$p_R = p_S \cdot \prod_{\substack{\forall C_{S^-} \neq C_S, \\ \mathrm{PER}(C_{S^-}, C_R) \neq 1}} (1 - p_{S^-}) \cdot \left(1 - \mathrm{PER}(C_S, C_R)\right) \tag{4-2}$$

This information will be used to evaluate packet receptions over the connections in the simplified multi-cell model of the PLC networks.

### 4.4.2 User Model in PLC Cell

In the multi-cell PLC model, the resource allocation sharing scheme reserves the resources for each cell based on the number of active users in the network. Each user may either have an active or inactive status. The user is in the active status when there is traffic to be transmitted from or to that user. According to [Chan00][GuLu07], a service is requested from a user following a Poisson process and the duration of each service (e.g. Voice or Video) is assumed to be negative exponentially distributed. Since the system runs with discrete time frame, the geometric distribution is used for modeling the duration of the services. For stationary situation, each service is assumed to be modeled by an ON-OFF model in which both ON and

OFF durations are geometrically distributed random variables, as has been introduced in the modeling of the bursty traffic source [LeeUn86].

Assumed that a user $U_i$ may use $K$ services and each service $k$ is modeled by an ON-OFF model with the transition probabilities $p_{on}^{ser(k)}$ and $p_{off}^{ser(k)}$ in a time frame. Probability that a service $k$ available in a time frame is

$$P_{serv}^{(k)} = \frac{p_{on}^{(k)}}{p_{on}^{(k)} + p_{off}^{(k)}} \tag{4-3}$$

And the probability that the service is not available in a time frame is

$$P_{no\_serv}^{(k)} = 1 - P_{serv}^{(k)} \tag{4-4}$$

Therefore, probability that there is no service in use, implying that the user is inactive, is

$$P_{no\_serv}^{(U_i)} = \prod_{k=1}^{K} (1 - P_{serv}^{(k)}) \tag{4-5}$$

and probability that the is at least one service available, implying that the user is active, is

$$P_{serv}^{(U_i)} = 1 - \prod_{k=1}^{K} (1 - P_{serv}^{(k)}) \tag{4-6}$$

From (4-5) and (4-6), probability that a user $U_i$ has inactive status for $x$ time frames is

$$P(Y = x) = (1 - P_{serv}^{(U_i)})^x P_{serv}^{(U_i)} \tag{4-7}$$

and probability that a user $U_i$ has active status for $x$ time frames is

$$P(X = x) = (1 - P_{no\_serv}^{(U_i)})^x P_{no\_serv}^{(U_i)} \tag{4-8}$$

(4-7) and (4-8) are probability mass functions of two geometric distributions. Therefore, changing a user's status can be modeled by an on-off model and both on and off durations are random variables which are geometrically distributed.

## 4.5 Types of Services in Single PLC Cell and PLC Link

### 4.5.1 Types of Services inside a PLC Cell

Based on different specifications and service providers, each PLC cell may use a specific PLC technology. However, each user may request one or several services. Each service is modeled by the type of service. In this investigation, four service classes with two types of services are studies, hard real-time and elastic services.

The service model refers to the service class definitions in [Opera06]: a service provider can provide different types of services, not only the hard real-time constrained services like VoIP (Voice over Internet Protocol) or video conferencing, but also the so-called elastic services, in which the perceptive quality of service does not strongly depend on the provided allocated bandwidth. In order to evaluate the QoS of the services, a so-called benefit function is defined to calculate the utilization of the granted bandwidth. The benefit function evaluates the usefulness of the allocated bandwidth for each granted service. Fig. 4-11 shows an example of benefit functions of an elastic service and a hard real-time service. The benefit of the real-time traffic is defined by a sigmoid function or a jump function while the benefit of the elastic traffic is defined by a concave function [Ozda08]. For a real-time service, if resources are allocated lower than the given threshold, the service may not be useable. The service is only useable, which results in a value of benefit, when the allocated resources are equal or greater than the threshold. For an elastic service, however, the service can be used with a small amount of resources. The benefit increases when the amount of allocated resources is higher. With too many allocated resources, the service may already be satisfied thus the benefit may not be further increased. One objective of the resource allocation method is to achieve the maximum benefit values of all granted traffic flows.

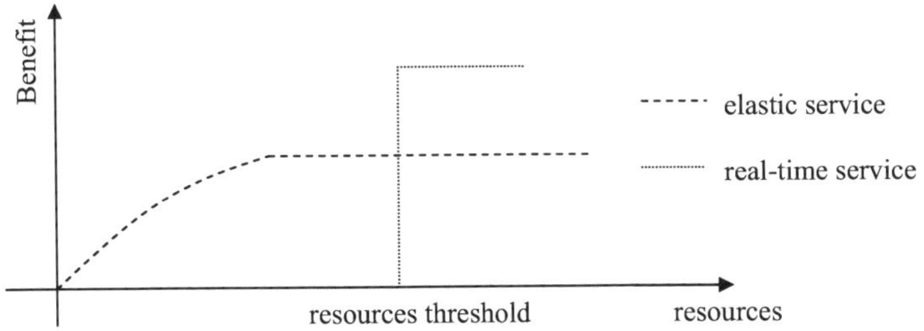


*Figure 4-11: Comparison of benefit functions for different type of services*

In this thesis, we assume that each user may run four types of services with QoS guarantee. The connection admission control (CAC) module decides if a connection is accepted or not. The decision of the CAC takes place at the beginning of each scheduling period. A scheduling period is the duration between two consecutive schedules. If a real-time service is accepted, an amount of transmission resources which is enough for the maximum traffic request is reserved for it. In case the real-time service is variable bit rate traffic, the unused reserved resources during the scheduling period are assigned to the accepted elastic services.

### 4.5.2 Model for PLC Link between Network Devices inside a PLC Cell

Depending on the access schemes, a link between two network devices contains of one or several frequency bands. Each frequency band has an individual transmission data rate over the same power line. The data rate of each frequency band is a random variable, generated in accordance with the parameters presented in Chapter 7.

## 4.6 Transmission Resource Model

We assume that a PLC link uses the Orthogonal Frequency Division Multiplexing (OFDM) modulation scheme. For an OFDM system, the smallest transmission unit is formed by a combination of a carrier and the duration of an OFDM symbol. The duration of an OFDM symbol and the number of available carriers of a PLC system depend on the utilized technologies.

For allocation of the transmission resources between the cells, a collaborative algorithm has to be performed at each base station of each cell. The smaller the granularity of the available transmission resource, the larger the signaling exchanged between the cells. Therefore, several carriers and OFDM symbols are grouped into resource units, called channels. Certain resource units are reserved for certain PLC cells. Depending on different PLC specifications, the resource unit is defined by a combination of the carriers and OFDM symbols in a specific method.

| *Name/Symbol* | Value | *Name/Symbol* | Value |
|---|---|---|---|
| $f_{min}$ | 1MHz | Number of frequency carriers | 1600 |
| $f_{Max}$ | 30MHz | Duration of an OFDM Symbol | 71µs |

*Table 4-1: System parameters of the transmission resource*

In accordance with [Opera06] and [HomePa07], each cell may be owned by a certain service provider; therefore, each cell can utilize different PHY layers having different specifications. Each specification utilizes the frequency band in accordance with different modulation methods, e.g. one utilizes the fast Fourier transform (FFT)-OFDM while another uses wavelet transform (Wavelet)-OFDM [AbadTo05]. System parameters of the PLC network to be used for evaluation of the proposed allocation protocol are presented in Tab. 4-1. It is assumed that the PLC system uses the frequency band from 1MHz to 30MHz. The available frequency band consists of 1600 carriers. Some carriers may be suppressed (notched) due to regulations, e.g. for radio amateur bands. The duration of an OFDM symbol is 71µs.

Our idea of the organization of resource units inside a time frame is presented in Fig. 4-12. The multiple access scheme is organized in the form of a hybrid of OFDMA and TDMA. The smaller number of resource units also requires larger overhead for identifying a certain resource unit. Overall, each resource unit is an element of a four dimensional resource allocation matrix: aside from the indexing in the frequency band (group of carriers) and time slot (group of the OFDM symbols), a resource unit carries two other parameters: user and service.

Overall, the four dimensions of each resource element are

- Group of OFDM carriers,
- Time slot index in every time frame,
- Transmission capacity of user in that frequency band given by the modulation index, and
- Type of service carried in that resource unit, e.g. real-time or elastic traffic.

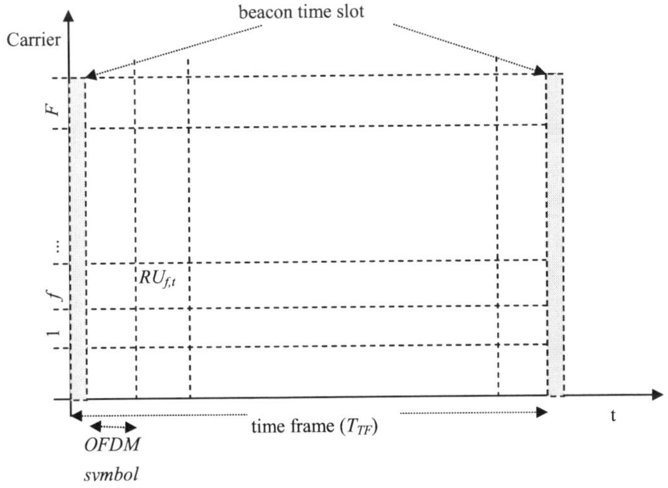


*Figure 4-12: Resource organization in a time frame*

The parameters used for the evaluation of the proposed protocol are derived from the [Opera06] specification and presented in Tab. 4-2. A smallest possible resource unit ($RU_{f,t}$) is formed by a carrier and an OFDM symbol, it carries up to 10bits. However, a resource unit when using a TDMA based MAC layer in accordance with [Opera06] is the combination of an OFDM symbol with the entire frequency spectrum. The time axis is subdivided into time frames (TF). For supporting time sensitive services that require short network access delays, a time frame is divided into eight sub time frame (sTF). The overhead used to identify the allocated resource unit for each user and service has to be considered when evaluating the efficiency of the resource allocation method.

| *Name/Symbol* | Value | *Name/Symbol* | Value |
|---|---|---|---|
| *Time Frame ($T_{TF}$) duration* | 240ms | *bits per carrier in a OFDM symbol* | up to 10bit |
| *Sub Time Frame ($T_{sTF}$) duration* | 30ms | *OFDM symbols per sub Time Frame* | 422 |

*Table 4-2: Resource parameters for the modeling*

# Chapter 5 Hierarchical Resource Allocation Model

This chapter presents a proposal for the transmission resource sharing architecture for PLC networks. The available transmission resources are firstly shared between the PLC cells within a PLC network. From the point of view of the network organization, the transmission resources available have to be organized in order to be shared by all the users in the PLC cells. Secondly, the reserved transmission resources are shared between the network devices under control of a base station. A PLC cell is configured as a master-slave system, thus all traffic between the subscribers in a cell and the backbone network is controlled and transmitted through the base station of the cell. Our new hierarchical transmission resource allocation model, including all the functionalities and procedures taking the PLC techniques, network organizations, and available transmission resources into account, is presented.

## 5.1 Two Levels of Transmission Resource Sharing Model

### 5.1.1 The Approach of the Two Level Model

Each BS in a PLC cell is the network device to perform the resource organization for two sides: one side is to obtain the reserved resources for its cell and the other is to manage and organize the allocated resources for its own users. Therefore, the resource allocation problem is divided into two sub-problems which form the so-called two level hierarchical transmission resource allocation model. Dividing the entire resource allocation problem into two sub-problems matches the PLC network model in which the relations between PLC network devices are logically organized on two levels: between PLC cells (inter-cell level) and between users in a PLC cell (intra-cell level).

The inter-cell level (upper level) consists of the task of sharing the available transmission resources between the cells. After this task is completed, each cell will have reserved a certain amount of transmission resources for transmissions within itself. In the intra-cell level (lower level), the allocated resources inside a cell have to be split into smaller portions to support the transmissions between all the network devices in that cell. Fig. 5-1 presents the two level transmission resource sharing approach and corresponding input and output of each level.

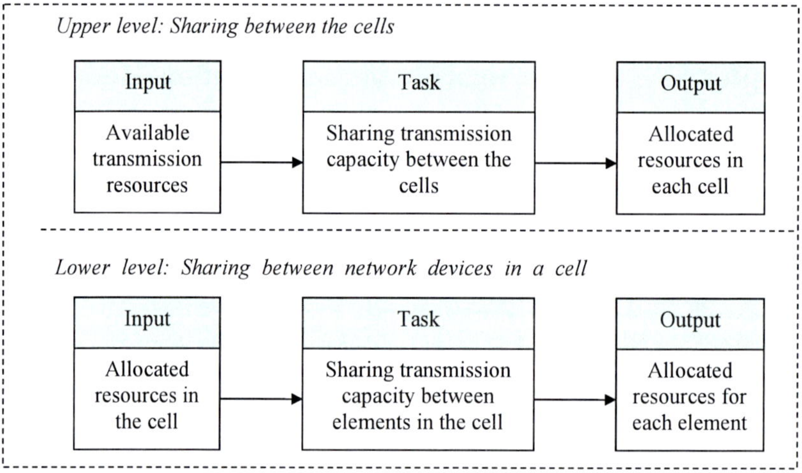


*Figure 5-1: Tasks in the two-level model*

The entire resource allocation problem is divided into two sub-problems following the relationship between PLC devices, that may also logically organized into two levels, between PLC cells and between users in a cell, as presented in Fig. 5-2.

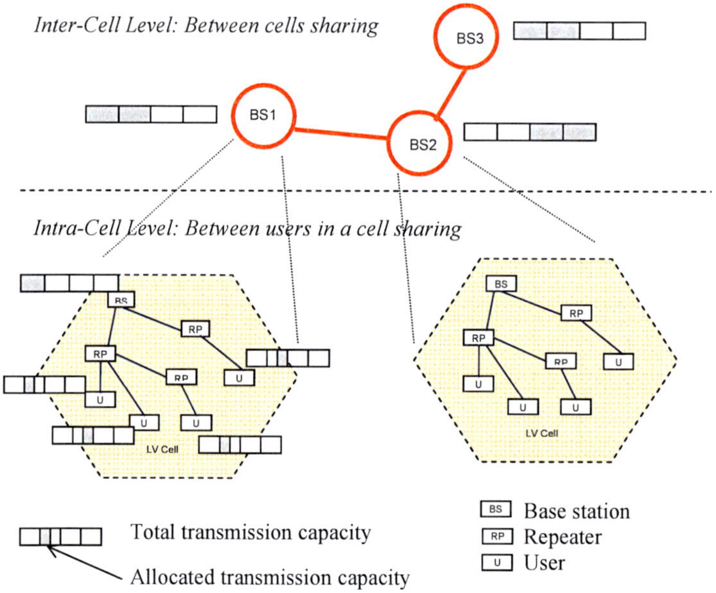


*Figure 5-2: Relationship between the levels of the model*

Generally, this approach has already been used in several cellular communication techniques e.g. GSM [AnanSr03], UMTS [AmalCa03]. It is also applied for PLC, in which the cells' organizations and the channel allocations are optimized with global information about the PLC network, during the network design period [Haid08]. For resource sharing between cells, these algorithms allocate a fixed number of channels to the cells or require a central decision unit. However, for the specific scenarios of a PLC network, in which a control center is lacking, a transmission-sharing scheme has to be designed with a distributed approach. Additionally, due to the variation in the transmission requests from each cell, the allocation method has to shift the transmission resources between the cells dynamically. On the intra-cell level, the transmission resource sharing strategies allocate the resources to each user with the aim to achieve highly efficient utilization of the bandwidth based on different services with different quality of service (QoS) requirements.

### 5.1.2 Resource Sharing in Inter-Cell and Intra-Cell Levels

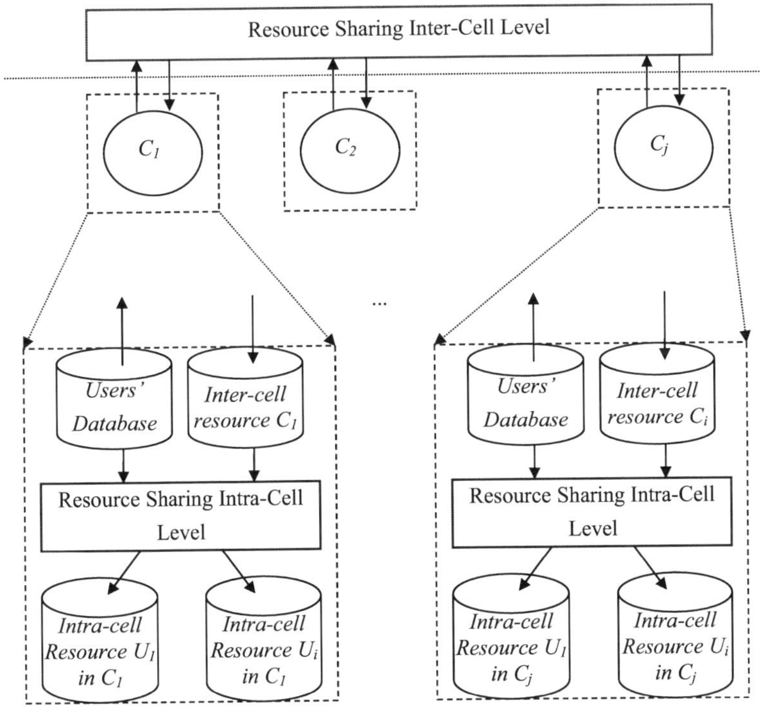


*Figure 5-3: Relationship between the resource sharing in two levels*

The relationship between the two levels of the resource sharing tasks is depicted in Fig. 5-3. For the inter-cell level sharing scheme, the inputs are the traffic request status of each cell and the relation between the cells. The outputs are the so-called inter-cell resources for each cell. These inter-cell resources are then used as the input for the intra-cell resource sharing calculation. The scheduler of each cell has to determine the intra-cell resources for each user in its cell. One example showing the resource variations in two cells $C_1$ and $C_2$ and a user $U_1$ inside $C_1$ is presented in Fig. 5-4.

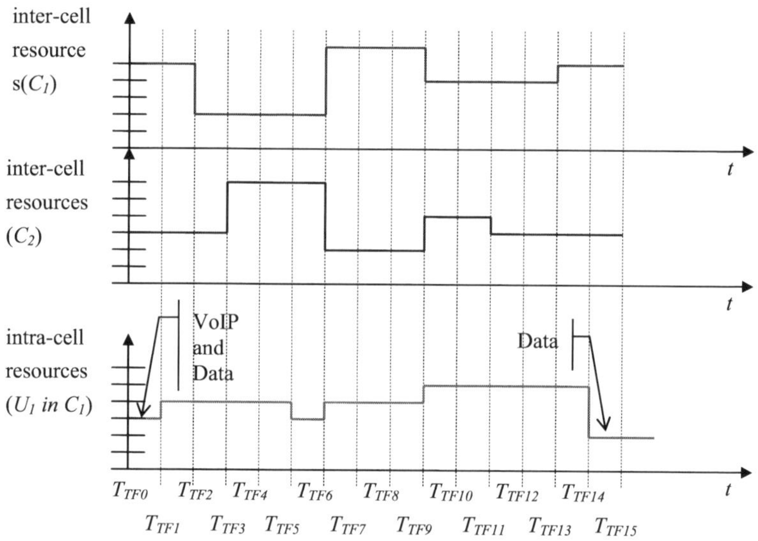


*Figure 5-4: Variation of resources after inter- and intra-cell resource allocations*

It is assumed that $C_1$ and $C_2$ operate synchronously and the system time unit is the time frame (*TF*). After $T_{TF0}$, the number of users in cell $C_1$ decreases, so that at $T_{TF1}$, the $C_1$ can grant more resources to its users. As a result from its scheduler, it provides more resources for its users. The user $U_1$, which at that moment uses VoIP and data transmission, obtains more resources for its traffic. By comparing with $C_2$, $C_1$ releases resources to $C_2$ at $T_{TF2}$. Because of the transmission between the cells, the released resources can be seized by $C_2$ in the next or several time frames later: in this example, it is one time frame later ($T_{TF3}$).

At $T_{TF5}$, due to additional activation of another user in cell $C_1$, the amount of resources of user $U_1$ is reduced until $T_{TF6}$ when cell $C_1$ receives resources again from cell $C_2$. The resources of user $U_1$ can also be reduced, because its resource requirement reduces, e.g., at the point of time $T_{TF14}$ after its VoIP is terminated. During the time in which the inter-cell resources do not adapt to the change of the intra-cell's traffic request, e.g., between $T_{TF5}$ and $T_{TF6}$, the scheduler in each cell has to arrange the available resources appropriately with regards to the current traffic request. Overall, the intra-cell resource allocation occurs more frequently than the inter-cell resource scheduling.

### 5.1.3 Design Requirements

There are technical requirements that have to be considered when applying the proposed two-level resource sharing design. The first is the interference and transmissions between the cells. The interference between the cells is discussed in section 5.2.1. The interference is considered when arranging the exchange of information between the cells without additional wiring. The organization of the information and transmission between the cells is presented in section 6.2. Each PLC cell is planned to be a master-slave system in which no further bandwidth reuse can be applied.

Secondly, the compatibilities of the equipment have to be considered. In reality, the equipment may be different in terms of the version, manufacturer, standards, but they are not allowed to inhibit the transmissions of the others. In other words, this equipment has to adhere to the common rule of coexistence.

## 5.2 Inter-Cell Transmission Resource Sharing

Due to interference between adjacent cells, both PLC specifications, [Opera06] and [HomePl06], propose to subdivide the transmission resources into smaller units to be distributed into the cells. However, these specifications are only aimed to deal with the co-existence between access cells and inhome cells, following a fixed allocation scheme. This does not address the dynamic traffic request from different cells in the overall PLC network. Previous studies have also proposed to allocate the transmission resources to the cells in the PLC network by utilizing an optimization problem solution. To solve these optimization problems, however, it requires full knowledge about the traffic requirement throughout the entire network. Therefore, the problem can only be solved during the phase of network design, e.g., using heuristic algorithms for channel allocation problems, see [Haid07] and [Haid08]. The results are therefore a fixed channel allocation solution, meaning that each cell is allocated a number of fixed channels and these allocated channels are not altered during run-time. This kind of solution leads to unfairness between users in different cells, when the traffic demand from users varies or the cell structure is changed.

On the other hand, there are several studies which have investigated to solve the channel allocation problem dynamically, but most of them were originally proposed for the GSM cellular system [FranNi97] [LoiWi99] [LozaDo02]. These available dynamic channel allocation methods, however, cannot be applied to the PLC directly, due to the differences between the systems. The main reason is that it is not feasible to organize a central unit for controlling the complete PLC network when the network is composed of several clusters operated by different service providers. A PLC user does not move and a private inhome PLC cell may even be organized as an intranet without an Internet connection. Based on these reasons, the re-organization of the PLC users between cells is not always possible, similar to the hand-over procedure in the case of the GSM system. Therefore, we propose a novel negotiation strategy to be applied to every base station (BS) in the PLC network. The strategy allows all BSs to co-operate and to share the transmission resources dynamically during run-time and in a distributed fashion. This strategy allows every cell to maximize the reserved channels, while minimizing the channel interferences, and maintaining the fairness between users in different cells. The sharing methods have to take several network parameters into account, e.g., size and resource requirement from each cell, interferences between cells, as well as the number of available channels. For our investigation, the reference networks used are based on the study of the electricity networks from the European OPERA project [Opera06], as presented in Chapter 4.

### 5.2.1 Interferences

Many types of interferences in powerline communications have been intensively studied, see [Maru97], [LaudSu99], and [Dost01]. Generally, when the communication signal is transmitted over the powerline, it can interfere with and be interfered by other signals, e.g., from radio signals or from the other PLC transmissions. Since the effect of the noise from the foreign systems is not avoidable, this investigation is interested in reducing the interferences caused by the PLC transmissions themselves. There are two kinds of interferences between the transmitting signals over the PLC networks, namely in-line interference and in-space interference [Dost01][Haid07]. With the PLC cellular concept, the concern is to reduce the possibility of interferences between the transmissions in adjacent cells. Figure 5-5 demonstrates the potential in-line and in-space interferences caused by the transmissions in two neighbor cells in a selected segment of a PLC network.

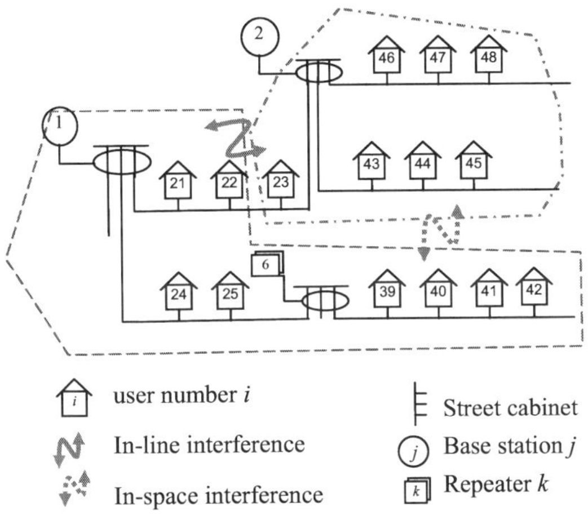


*Figure 5-5: Types of interferences between neighboring cells*

- In-line interferences occur when there is a power line used by both cells. E.g. in the line between two street cabinets in which BS1 and BS2 are installed. These two cells become neighbors. In this case, the transmission from a network device to one cell will be disturbed by the neighbor's if they use the same frequency band at the same time. It is a kind of collision between transmission signals. Therefore, these network devices from both the cells have to use different frequency bands or time slots for their respective transmissions.
- In-space interferences occur when two power lines carrying PLC signals are physically installed near each other. The cause of the interference is the electro-magnetic field that one powerline cable creates over the air of another cable. In this case, both of the powerline cables passively work as antennas emitting and receiving the signal to and from the other. The in-space interference, which depends on the wiring of the network, causes fewer problems than the in-line interference when considering the actual locations of the neighboring cells [Dost01][Haid07].

The avoidance of the interference has to be organized by the master station when it does scheduling. If there is potential interference in the link to a station, the transmission in this link will be organized in a different time slot or frequency band of the interfering cell. The other links, which are not affected by any interference from the neighbor cells, can freely use all available time slots and frequency bands. In the worst case, if all the links in a cell are interfering with the neighbor cells, all the transmissions have to be scheduled in the time slots and frequency bands that are not used by the neighbor cells.

### 5.2.2 Channel Organization

We define a channel as a unit of transmission resources which can be allocated to a cell. The PLC systems are assumed to use Orthogonal Frequency Division Multiplexing (OFDM) modulation as specified in the PLC specifications [Opera06][HomePa07][D28OP07]. Generally, a channel can be defined as a group of carriers, a repetition of a time slot in each time frame, or a hybrid of time slot and frequency carriers. Advantages and disadvantages of the channel organization are analyzed.

- A channel may be defined as repetition of time slots in each time frame, as presented in Fig. 5-6. In this kind of channel organization, each time frame is divided into time slots. Each time slot may consist of several OFDM symbols. The advantage of this channel organization is that the entire frequency band is assigned to a cell at an instant of time to avoid frequency selective noise or notching in some specifically selected frequencies. Suppressions of the signal in some carriers drop the transmission capacity of some specific frequency bands. However, when a cell has reserved some time slots that are not continuous in a time frame, the fragmentation of the transmission time increases the complexity of the later scheduling for users inside a cell. This may also reduce the efficiency of the resource utilization because multiple transmissions in different time slots of a user require additional overheads. This system also requires time slot synchronization between the neighbour cells. Based on accuracy of the synchronization, guard interval between the time slots may be necessary.

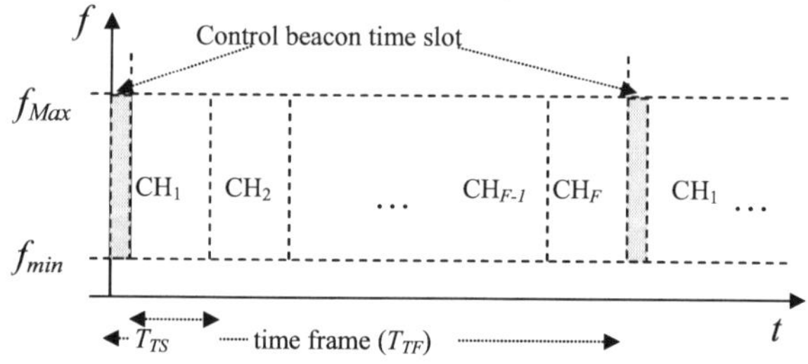


*Figure 5-6: Channels as time slots*

- A channel may be defined as group of carriers, as presented in Fig. 5-7. Contradictory to the previous case, this channel organization allows each cell to be reserved one or several groups of carriers in entire time frame. With the utilization of the whole time frame, realization of sensitive time service which requires short transmission delay is feasible. The access to the entire time frame requires a lower calculation complexity for the scheduling, since each cell is allowed a full flexibility to access the medium during run-time. For supporting different PLC specifications, each cell can utilize a particular modulation scheme inside the reserved frequency channels. Main disadvantage of this channel organization is that it is too complex to include channel quality of each channel in all the cells into the inter-cell channel allocation process. This is because a channel may be scheduled in a cell in several intra-cell connections, each having a different channel quality.

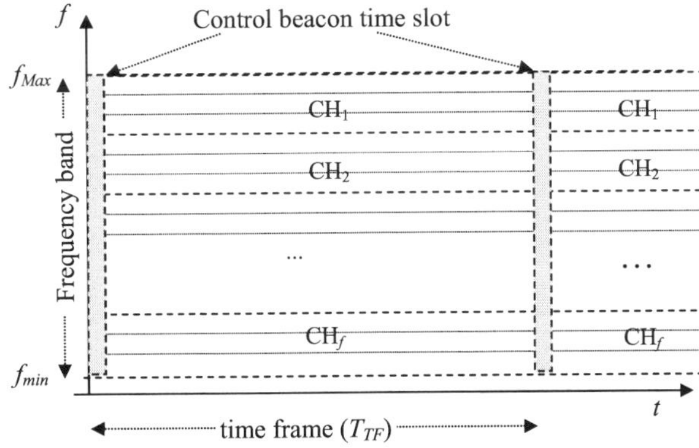


*Figure 5-7: Channels as groups of carriers*

- A channel is defined as group of time slots and carriers. This channel organization has the advantages and disadvantages of the both previous channel organizations. It allows a full flexibility in both time and frequency dimensions for the resource allocation problem. However, it also increases the complexity of the scheduler inside each cell which may not applicable for the PLC system in reality.

In this thesis, a channel is defined as a group of carriers. The organization of the channel as inter-cell resource units in the time frame is presented in Fig. 5-7. If a channel consists of many carriers, there are only few channels remaining for sharing between the cells. This may lead to a low flexibility when allocating the channels to the cells. On the other hand, if there are a high number of channels, because a channel is a small group of carriers, the information which is used for indexing the channels and which has to be exchanged between the cells is large.

Considering a multi-cell PLC network which has $N_{All}^{Carrier}$ available carriers: if each channel is a group of $N_{1Channel}^{Carrier}$, the number of channels $F$ is

$$F = \frac{N_{All}^{Carrier}}{N_{1Channel}^{Carrier}} \tag{4-9}$$

Our proposed solution operates with channels as inter-cell resource; therefore it is independent from the divisions of the frequency and time domains.

## 5.3 Intra-Cell Transmission Resources Sharing

Each specific PLC user logically connects to a base station and is controlled by that base station in a particular cell. Generally, a PLC cell works as a master-slave system, where the base station is the master and the repeaters and users are slaves. In this network model only the master is connected to the backbone network to provide network access to the all customers inside its cell.

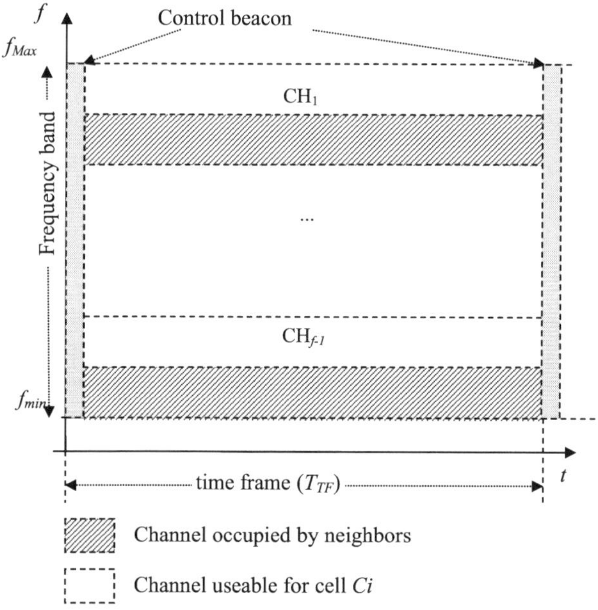


*Figure 5-8: Transmission resources for a cell in a time frame*

The available transmission resources for a cell are the output of the sharing procedure in the inter-cell level. Fig. 5-8 presents the organization of transmission resources in a cell. Since the service requests and channel qualities are stochastic, the connection admission control and the scheduler have to dynamically adapt to these changes. Another requirement is that the time to perform the resource allocation solution must be short, in accordance with the requirements from the services in the network.

The transmission capacity of a user is estimated by the transmission data rate between that user and the base station and the transmission time. Since different users have different transmission data rates to the base station, the scheduler allows each user a different amount of transmission times.

### 5.3.1 Network Organization

The logical network structure inside a PLC cell depends on the location of the nodes and channel qualities between the network devices. Therefore, the logical PLC cell structure can vary during the run-time, based on the variations of the users' statuses and the channel qualities. In reality, the network can be set up and maintained by means of the best master selection protocol presented in Appendix A.1.5.2. For the investigation of the inter-cell resource sharing, the logical network structure is used as the input for the resource management algorithms, in which the decision on resource allocation to particular connections and services is made. Therefore, for the evaluation of the resource sharing protocol, the network structure is assumed to be constant during the simulation time. Only traffic requests and channel qualities of the connections are varied during the simulation run-time.

### 5.3.2 Resource Management and Sharing

The resource management module has to manage the resources usage by each user in the network. When there are enough transmission resources for all the service requests, the resource management simply grants the sufficient amount of resources to all the requests. However, in case the available resources are smaller than the sum of all the requests, as depicted in Fig. 5-9, the resource manager has to decide how much of the resources to grant to a particular service for a specific user. The objectives of resource allocation mechanisms are to achieve a high utilization of the allocated transmission resources while to maintain fairness between the users.

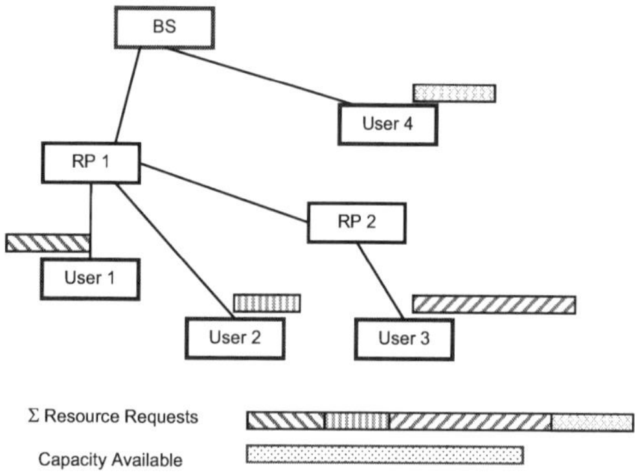


*Figure 5-9: Resource requirements in a network*

The resource management strategies are proposed to be used with different PHY layers. For the design of the strategies, functionalities of the MAC protocol are necessary. The investigation is performed under the assumption that the MAC/PHY layers are specified according to the OPERA specifications [Opera06]. Based on the data management in a base station depicted Fig. 3-1, a block diagram of the resource management is presented in Fig. 5-10. The databases to be stored and regularly updated in the base station are available transmission resources, current traffic, QoS profiles, channel quality, and user statuses. The functions of the resource management, connection admission control (CAC), traffic classification, buffer manager, and frame builder use these databases during their processes.

The main function of the resource management module is the traffic scheduler. The scheduler, based on the available resources, current traffic in the network, and the traffic request, determines the concrete transmission resources allocated to each connection. When a connection request arrives, the CAC module accepts or rejects the connection based on the decision on the allocated bandwidth from the scheduler. If the connection is accepted, the traffic is classified and stored in the appropriate queues, managed by the buffer management module. In accordance with the output from the scheduler, the frame builder takes data from corresponding queues to build the data frame to be sent into the network.

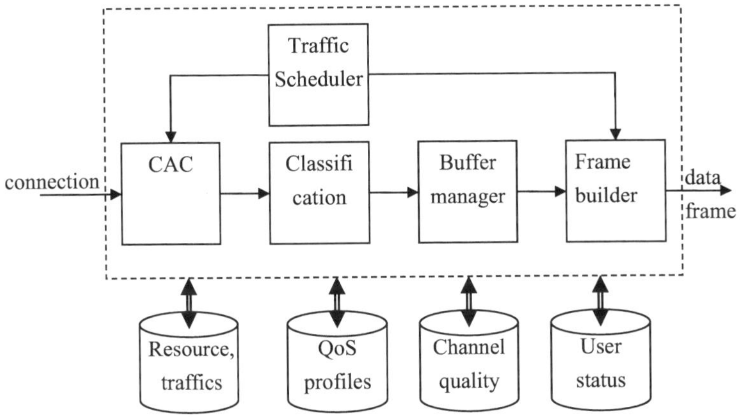


*Figure 5-10: Resource management in a base station*

The flow chart in Fig. 5-11 presents the algorithm for the CAC module. Each service class is predefined in the QoS profile database in terms of the QoS parameters, bandwidth, and latency. In order to guarantee a certain service, the base station has to reserve a predefined amount of transmission resources intended for this service class.

A request for a new connection can be initialized from both the uplink and downlink directions. If the request for the new connection comes in the downlink direction, meaning from base station to a user, the base station compares the resource requirement for the new connection with the available free resources. If there are enough resources, the connection is accepted with the required QoS parameters. Otherwise, the connection will be granted a lower QoS class or rejected. Some strategies can be applied in the decision when the network reaches its maximum transmission capacity, e.g. some of the lowest QoS service class connections may be reduced or dropped in order to gain enough bandwidth for a new and more important service. If the connection comes from a user in the uplink direction, a request for a new connection is sent to the base station. The process for connection admission is similar to the case of the downlink connection and the result is informed to the requested user. After receiving the decision from the base station, the user knows whether the connection was accepted with a certain QoS parameter or rejected.

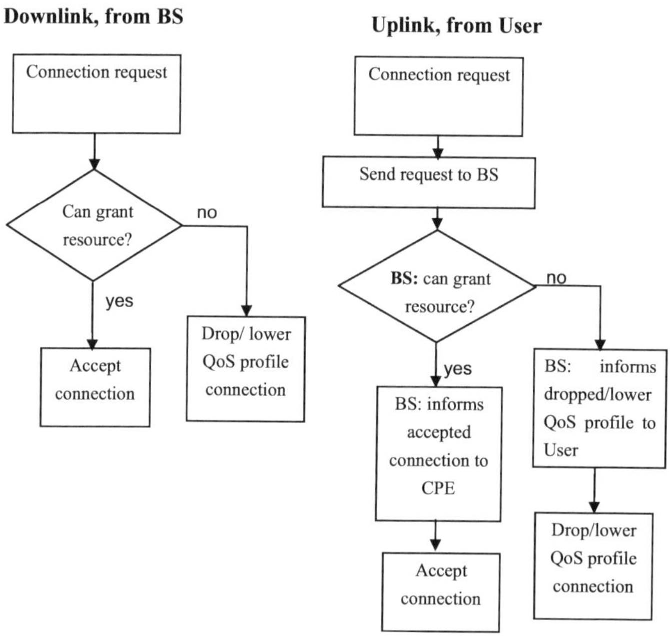


*Figure 5-11: Flow chart of the process of the CAC in both Downlink and Uplink*

### 5.3.3 Sharing Resources within PLC Cell

The limited transmission resources have to be shared by the PLC network devices, under the condition that the nodes in a cell are not allowed to use the same resources simultaneously. In general, the frequency band and the time frame are two dimensions that define a resource unit, as presented in Sec. 4.6. In accordance with [HrasHa04], the FDMA, TDMA, and their hybrid can be utilized as multiple access schemes for users in PLC networks.

- Using the Frequency Division Multiple Access (FDMA) scheme, the entire available frequency band is divided into several frequency segments. Each network node is allocated a certain segment for its transmission. This requires a mechanism to ensure that the interfering nodes do not use an overlapping frequency band. The transmissions from the nodes in this case are simultaneous. One disadvantage is that the transmission capacity of a channel may vary during run-time thus it is not possible to guarantee the QoS for constant bit rate streams.
- Utilizing the Time Division Multiple Access (TDMA) scheme, the time is subdivided into several time slots within a time frame. Each network node is given a certain time slot in every time frame for transmission using the entire available frequency

spectrum. The given time slot to a node does not overlap with the time slot of the interfering node. This scheme may cause a long delay to time sensitive services if a time frame is long.

- Utilizing the flexibilities from the resource allocation scheme mentioned above, a hybrid of the FDMA and TDMA can be used. Both frequency and time domain are divided into small segments. Similar to the allocation schemes above, each node is given a number of segments for its transmission.

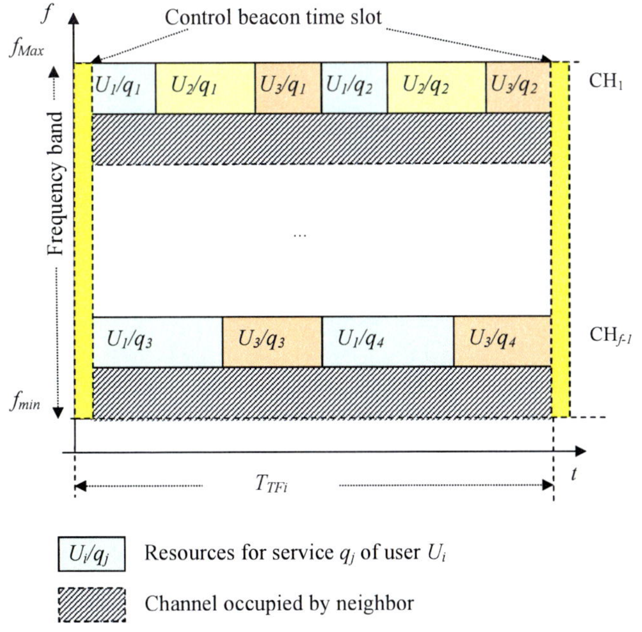


*Figure 5-12: Organization of resources in a PLC cell*

The investigation of the scheduling inside PLC cell is done in two parts. The first part deals with the TDMA access scheme, under the assumption that the MAC layer of the OPERA specifications [Opera06] is used. In the second part, the hybrid of the FDMA and TDMA is taken into account. Individual channel quality of each channel in each connection is considered by the scheduling process to improve overall system performance. Fig. 5-12 presents an example of the results from the scheduling process as way of organizing the resource allocation for all users in that cell.

# Chapter 6 Distributed Inter-Cell Resource Allocation Protocol (DICRA)

The proposed inter-cell resource allocation provides the allocation of the transmission resources to each PLC cell. In this chapter, the task of channel allocation means to reserve channels to each cell in the network. Based on the exchange of the resource utilization and traffic requests with neighboring cells, a cell adjusts its transmission capacity reservation accordingly. Methods for transmitting information between neighboring cells and criteria for the adjustment of the transmission resources are proposed and analyzed. The performance of the proposed scheme is evaluated based on the network reference models.

In this multi-cell PLC network, each PLC cell has one base station to provide communication services to all users. It is connected to the backbone network. The entire PLC network operates with limited transmission resources. When the cells are neighboring, the transmissions in a cell may produce interference to the transmissions of other cells. This means that neighboring cells cannot simultaneously utilize the same carriers. Another characteristic of the network is that the resource requirements from each cell are dynamically changing, e.g. users turn on or turn off during run-time. Therefore, a dynamic channel allocation protocol between the cells is necessary. This chapter proposes a channel allocation protocol to dynamically organize the limited transmission bandwidth between multiple PLC cells, which runs distributed fashion. The mechanisms are investigated to show their ability to improve the utilization of the overall transmission resources, while maintaining fairness between users and between the cells. The problem is summarized as follows:

- Given:
    - PLC network with multi-cell structure
    - A PLC cell can be an access or inhome cell
    - Each PLC cell has a specific owner / service provider
    - Traffic demand from each cell varies over the run-time
    - Limited number of available channels
- Objectives:
    - High reuse of channels
    - Maintain fairness between active users in different cells
    - Minimize interference

  - Minimize delay of channel reallocation (adaptation time)
- Requirements:
  - Dynamically adapt the channels to the changes of traffic demands
  - Adapt to the changes of the PLC cells and cells' structures during run-time
- Data and functions within the allocation protocol:
  - Channel status
  - Traffic demand description
  - Inter PLC cell connections
  - Local channel allocation mechanism based on channel information from neighboring cells
- Performance evaluation metrics:
  - Communication overhead for the protocol
  - Performance of the protocol with different PLC network structures
    - Channel reuse factor
    - Fairness between users
    - Interferences
    - Adaptation time (delay of channel reallocation) and stability of the protocol

In order to adapt to the characteristics of the PLC network, a common set of actions, which form a so-called Distributed Inter-Cell Resource Allocation (DICRA) protocol assigned to every base station (BS), is proposed. The protocol is based on exchanging the channel information between neighboring cells. The exchanged information is processed by each BS individually. Based on the internal status and the information received, each BS may decide if it has to release or seize channels.

## 6.1 Protocol Description

The negotiation strategy is based on the processing of the information exchanged between neighboring cells. At a specific point in time, a base station (BS) has to process the received information and then broadcast its knowledge about the channels using a so-called channel announcement message (CA-Msg). Based on the reception of CA-Msg from its neighbors, the BS has to release the interfering channels by reallocating channels or trying to seize other channels for its cell or to release channels of its neighboring cells.

### 6.1.1 Objectives

The protocol is designed with the previously mentioned objectives. The number of channels is limited; however, the total number of channels utilized in each cell can be increased by reusing of the channels in different cells in which using them in a cell does not interfere with transmissions in other cells. Avoiding using the same channels in two neighboring cells can eliminate potential interference between these cells.

There are different definitions of the fairness between the users in different PLC cells based on the scenarios of the PLC network. Considering two PLC cells in which each cell may interfere with the other, these two cells may be in one of the following situations:

- Case 1: Two access cells belong to the same service provider
- Case 2: Two access cells belong to different service providers
- Case 3: Two inhome cells belong to the same owner
- Case 4: Two inhome cells belong to different owners
- Case 5: The access cell belongs to a service provider and the inhome cell belong to a different owner

Because of different qualities of links between users and their master, each user may have a different transmission rate. In other words, each user may use a different transmission time to transmit a given data volume. For each situation, the arrangement of resources between cells to maintain one of the four fairness criteria: the transmission rate of each active user, transmission resources per cell, transmission resources per active user, or transmission rate per traffic request. These fairness criteria are analyzed as flows:

- Fairness of transmission rate per traffic request

To maintain this fairness, a cell is allocated resources to achieve a transmission rate proportional to the number of traffic requests in that cell. If two cells belong to different owners, as in case 2, case 4, or case 5, the fairness of transmission rate per traffic request is not suitable. It is because when users in a cell with low quality links (thus low data rates) request high transmission traffic, they are allocated a large amount of transmission time for guaranteeing the requested transmission rate. This method of resource allocation reduces the performance of the other cell of the other owner. In the case in which two cells belong to the same owner, as in case 1 and case 3, maintaining this fairness requires a fast exchange of the transmission resources between the cells because the total traffic requests from each cell varies frequently. Based on a change of the users' services, the stationary time of the total traffic requests is shorter than the stationary time of a user.

- Fairness of transmission resources per cell

Using this fairness criterion, the transmission resources are uniformly distributed into the cells. This criterion is introduced in [HomPa07] and analyzed in [GallKu09] with the so-called inter-PHY protocol (IPP), in which the transmission resources are defined in the time domain. If there are two cells, each is allocated half of the available transmission resources. However, this fairness criterion is not suitable for cases of the same owner as in case 1 and case 3, where the cells have different number of users. In the worst case, one cell with one user is allocated the same transmission duration as the other cell which has several users. For the other cases in which the cells belong to different owners, there is no request for shifting the transmission resources.

- Fairness of transmission rate per active user

To maintain this fairness, a cell is allocated resources to achieve a transmission rate proportional to the number active users. Similar to the situation when applying fairness of transmission rate per traffic request, when links with very low transmission bitrates are located in a cell, they request high transmission resources (i.e. a long transmission time or large frequency bandwidth) for guaranteeing the transmission rates. This means the amount of transmission resources allocated in the other cell, which belongs to the other owner as in case 2, case 4, and case 5, is reduced and thus not acceptable for the other owner.

- Fairness of transmission resources per active user

To maintain this fairness, a cell is allocated an amount of transmission resources proportional to the number of active users. By applying this fairness criterion, each cell is reserved a portion of the transmission resources based on the number of active users inside that cell. It is a balance between the fairness of transmission resources per cell and fairness of transmission rate. There are disadvantages of this fairness criterion in case different owners of the two cells, as in the case 2, case 4 and case 5, are reserved a different portion of the transmission resources because of the different number of active users in the other cell. However, the advantage of this criterion is that the low qualities of the links in a cell do not influence the performance of the other cell. Also when the transmission rate is not taken into account, the requests for the shifting the transmission duration between the cells also does not occur as often as when applying fairness of transmission rate.

The main advantages and disadvantages of maintaining each fairness criterion are presented in Tab. 6-1. The plus and the minus signs in the table represent advantages and disadvantages in each fairness criterion in each situation of the network, respectively.

| Fairness criterion \ Cell and owner | Same owner (case 1, case 3) | Same cell type, different owner (case 2, case 4) | Different cell type, different owner (case 5) |
|---|---|---|---|
| transmission rate per active user | +same transmission rate per user in different cells | -may reduce the overall performance | -may reduce the overall performance |
| transmission resources per active user | +same transmission resources per user in different cells | +transmission resources of a user in a cell do not depend on the links' qualities in the other cells | +transmission resources of a user in a cell do not depend on the links' qualities in the other cell<br>-same treatment for access users and inhome users |
| transmission resources per cell | -may not be fair between users in different cells, in terms of resources and rate | +fairness in the transmission resources of each owner | +fairness in the transmission resources of each owner |
| transmission rate per traffic request | -low performance of the user with fewer traffic request | -may reduce the network performance | -may reduce the network performance |

*Table 6-1: Comparison of the fairness criteria*

In accordance with the comparisons, the criterion "fairness" in terms of transmission resources per active user is a trade-off between the reservation of transmission resources per cell, transmission rate per users, and the interest of the cell owner. Therefore, fairness in terms of the allocated transmission resources per active user is one objective of the inter-cell resource allocation. The transmission resources are multiple of channels.

### 6.1.2 Performance Metrics

For evaluating the channel allocation problem, numerical performance metrics have to be defined. From the point of view of the service provider, the higher the sum of the used channels in all the cells in the network, the better. With high number of allocated channels thus potentially high transmission rate, a service provider may grant more services to the users. The objective may be to maximize the sum of the channels used in the network. The channel reuse factor is defined as the ratio of the sum of the channels used and the number of available channels. This factor also indicates the quality of the solution because it is independent from the actual number of the available channels. With the chosen fairness criterion as presented in Sec. 6.1.1, each cell is allocated the number of channels proportional to the number of active users. The number of interfering channels is also considered as one objective of the channel allocation solution.

#### 6.1.2.1 Channel Reuse Factor

The total number of channels can be higher than the number of available channels ($F$) because a channel can be used in parallel by two or more cells, if their transmissions do not interfere. The interference free transmissions in a same channel can be realized when the cells are far enough from each other. Since reusing a channel increases the overall network performance, we consider it as an objective and try to maximize this during the channel allocation procedure. As defined above, the set of $S$ cells is

$$\boldsymbol{C}=(C_1,..,C_S). \tag{6-1}$$

Each cell $C_i$ contains $N_i$ active users. The set of total available channels in the network is

$$\boldsymbol{F}=\{f_1,..,f_F\}. \tag{6-2}$$

Let $\boldsymbol{A}_{CA}$ be the channel allocation matrix of the entire PLC network:

$$\boldsymbol{A}_{CA}=\{a_{i,k}\}, \tag{6-3}$$

where $a_{i,k}$=1 or $a_{i,k}$ =0 indicates that the channel $f_k$ is used by the cell $C_i$ or not. By simply counting the channel allocation $a_{i,k}$ from $\boldsymbol{A}_{CA}$ for the cell $C_i$, we have the total number of channels seized by this cell is

$$B_{Ci} = \sum_{k=1}^{F} a_{i,k} \tag{6-4}$$

Therefore, the total number of channels used by all the cells $\boldsymbol{C}$ in the network is

$$B_{Ch} = \sum_{i=1}^{S} B_{C_i} = \sum_{i=1}^{S}\sum_{k=1}^{F} a_{i,k} \tag{6-5}$$

We define the channel-reuse factor $f_{CR}$ as the ratio of the actual number of channels used by all the cells to the number of available channels in $\boldsymbol{F}$, thus

$$f_{CR} = \frac{B_{Ch}}{F} \tag{6-6}$$

#### 6.1.2.2 Fairness between Channels per Active User in different Cells

The inter-cell fairness is defined under the assumption that there are no channels reused inside a cell. If reuse inside a cell is possible, e.g. because the MAC layer of the cell can utilize a channel for several users inside a cell, this does not effect the calculation of the inter-cell fairness.

For evaluation, we use the idea from Jain's fairness-index [JainCh84]. Let $M_i$ be the number of seized channels per active user in cell $C_i$. A cell which has only master active should always be ready to serve new users. Thus, it is allocated a number of channels as if it has one active user. The $M_i$ is calculated by

$$M_i = \begin{cases} B_{Ci} \text{ if } N_i = 0 \\ \frac{B_{Ci}}{N_i} \text{ if } N_i > 0 \end{cases} \tag{6-7}$$

where $N_i$ is the number of users in the cell $C_i$. The set of users in the network with $S$ cells is

$$\boldsymbol{N_U}=\{N_1,..,N_S\}. \tag{6-8}$$

We attempt to maintain the same $M$ value for all the cells. Therefore, the fairness-index of a channel allocation is calculated by

$$F_{Ch} = \frac{\left(\sum_{i=1}^{S} M_i\right)^2}{S \cdot \sum_{i=1}^{S} (M_i)^2} \tag{6-9}$$

The value of the fairness-index is 1 if all allocations of channels per number of users in all cells $C_i$ are exactly equal and decreases to less than 1 when deviating from this situation.

#### 6.1.2.3 Interferences

As it is described in Sec. 5.2.1, only parallel transmissions in neighboring cells may interfere. Therefore, each cell only knows the number of channels which interfere with its set of allocated channels. Let $\boldsymbol{I_{in}}$ be the interference matrix in which an element $i_{i,j}$ has value of 1 or 0 indicating that a transmission in the same channel in the cell $C_i$ and cell $C_j$ is interfering or not, respectively.

$$\boldsymbol{I_{in}} = \{ i_{i,j} \} \tag{6-10}$$

The number of interfering channels each cell experiences is counted by examining each channel used by its neighbor cells. In cell $C_i$, if a transmission using channel $f_k$ interferes its neighbor cells or not is determined by:

$$I_{Ch}(i,k) = a_{i,k} \bigvee_{\substack{j=1 \\ j \neq i}}^{S} (i_{i,j} \cdot a_{j,k}) \qquad (6\text{-}11)$$

$\vee x_i$ is the logic OR operator for the elements $x_i$ ( $x_i \in \{0,1\}$ ). In (6-11):

$$I_{Ch}(i,k) = \begin{bmatrix} 1 & \text{if at least one neighbor cell allocates the same channel } f_k \\ 0 & \text{otherwise} \end{bmatrix} \qquad (6\text{-}12)$$

For evaluation purposes, the number of interfered channels is totaled for every cell. The total number of interfering channels used in the whole network is

$$I_{Ch} = \sum_{i=1}^{S} \sum_{k=1}^{F} I_{Ch}(i,k) = \sum_{i=1}^{S} \sum_{k=1}^{F} a_{i,k} \bigvee_{\substack{j=1 \\ j \neq i}}^{S} (i_{i,j} \cdot a_{j,k}) \qquad (6\text{-}13)$$

The number of interfering channels is used for evaluating the channel allocation solution. Transmission in these channels may be interfered by the transmissions in the same channels in the neighbor cells.

#### 6.1.2.4 Adaptability of the Proposed Protocol

The adaptability of the DICRA protocol is described by the reaction on channel reallocation after changes of the resource requests from the neighboring cells. The decision of each cell is however based on the channel information that it receives from its neighbors. If the status information of cell $C_j$ in cell $C_i$ is invalid, e.g. the channel information from the neighbor $C_j$ has already changed while the cell $C_i$ does not know about it, then cell $C_i$ uses outdated channel information from cell $C_j$ to determine the channel allocation. Therefore, we evaluate the adaptability of the protocol indirectly by evaluating the duration for delivery of the information between two neighbor cells, in number of time frames, denoted as $D_{j \to i}$ for the duration for delivery from cell $C_j$ to cell $C_i$.

As assumed in Sec. 5.2.1, two cells are neighboring if their transmissions may interfere, because they share a power line (in-line interference). Therefore, a connection is present between two neighboring cells. In this context, a connection between cell $C_i$ and cell $C_j$ implies the end-to-end connection between two BSs of two cells. With this assumption, the variable $i_{i,j}$ equal to 1 or 0 is also used to indicate whether a connection between cell $C_j$ and cell $C_i$ is present or not.

Let $n^{conn}$ be the number of end-to-end connections between all pairs BSs,

$$n^{conn} = \sum_{i=1}^{S-1} \sum_{j=i+1}^{S} i_{i,j} \qquad (6\text{-}14)$$

Because the delivery time from cell $C_j$ to cell $C_i$ may be different from that of cell $C_i$ to cell $C_j$, the average delivery time is the average of the delivery times in all end-to-end connections between all pairs of BSs:

$$\overline{D_{Ch}} = \frac{1}{n^{conn}} \sum_{i=1}^{S-1} \sum_{j=i+1}^{S} i_{i,j} \frac{D_{j \to i} + D_{i \to j}}{2} \tag{6-15}$$

In (6-15), the element $i_{i,j} \frac{D_{j \to i} + D_{i \to j}}{2}$ indicates that the delivery time is only calculated in case a connection between cell $C_i$ and cell $C_j$ is available.

### 6.1.3 Problem Formulation

Since we consider a PLC network consisting of $S$ cells, from $C_1$ to $C_S$, we have a set of cells $\boldsymbol{C}=\{C_1,..,C_S\}$. At a point in time, each cell $C_i$ contains $N_i$ active users. We assume that there are a total of $F$ channels to allocate to these cells, and the set of channels is $\boldsymbol{F}=\{F_1,..,F_F\}$.

Because of the limited amount of available transmission resources, defined here as channels, the channels should be reused by the cells in which using these channels does not interfere with transmissions in the other cells. Therefore, one objective of the proposed DICRA protocol is to maximize the total number of allocated channels calculated in (6-5). Maximizing this value also means maximizing the channel reuse factor calculated in (6-6). However, only maximizing the channel reuse leads to unfairness between the cells, since more channels tend to be reserved for the cells in which channels can be highly reused. The cell in which the channels cannot be reused is reserved only for a low number of channels or possibly no channels. As a result, the second objective of the protocol is to maintain fairness between the cells. The fairness index is calculated in (6-9).

The protocol is designed to avoid the situation in which neighboring cells use the same channels. The number of interfering channels is monitored and calculated by (6-13) to evaluate the effective number of channels, $B_{Ei}$ of a cell $C_i$, defined as the number of channels in use minus the number of interfering channels. The delivery time of the channel information calculated in (6-15) is also minimized by the appropriate transmission mechanism.

The channel allocation problem is formulated as a multi-objectives optimization problem:

*maximize* $f_{CR}$ in (6-6)

*maximize* $F_{Ch}$ in (6-9)

*minimize* $I_{Ch}$ in (6-13)

*minimize* $\overline{D_{Ch}}$ in (6-15)

with the parameters and variables defined in

(6-1) to (6-5), (6-7) to (6-8), and (6-10) to (6-14).

### 6.1.4 Principle of the DICRA Protocol

If all the network information is known by a centralized station, the problem can be solved by one of the multi-objective optimization solvers [FonsFl93] [DebPr02]. The problem can be converted into a single-objective optimization problem, e.g. by means of a weight aggregating function in which each objective is given a "weight" based on its influence to the result [SumaKu06]. By doing so, the problem is solved by one of the single objective optimization solvers [SumaKu06][IshiNo06]. However, due to a lack of such centralized station, the problem has to be solved locally by each base station, based only on local channel information received from its neighbors. A protocol to achieve the objectives presented in 6.1.3 is proposed.

The process of the distributed dynamic channel allocation is designed based on the characteristics of the PLC networks that it is feasible to organize the exchange of information between the base stations of the different cells. The protocol allows a base station to determine on the channel allocation by comparing its actual traffic status, including traffic request and channel reservation, and the traffic statuses of its neighboring cells. The exchange of the traffic status is realized over the shared or connected cables between neighboring PLC cells. The principle of the protocol follows the scheme of the distributed dynamic channel allocation as presented in section 3.6.2.3.

- To maximize the channel reuse factor $f_{CR}$, each cell attempts to seize as many channels as possible. This objective is conflicting with the objectives to maximize the fairness-index and to minimize the number of interferences. Trying to maximize only the channel reuse factor may result in a low fairness index because of the tendency to assign more channels to a cell which has a higher possibility of being reused. Allocating more channels to neighboring cells increases the potential of using the same channels in these cells that may result in a higher number of interferences.
- To maximize the fairness index $F_{Ch}$, each cell releases channels if its $M_i$ value is higher than its neighbors. If a channel, which is released by a cell with a higher $M$, is seized by another cell, $f_{CR}$ does not change but the $F_{Ch}$ increases.
- To minimize the number of interferences, $I_{Ch}$, each cell immediately releases a channel if it is detected that the channel is used by its neighbor cell.
- To minimize the duration of the transmissions between the cells, $\overline{D_{Ch}}$, the transmission of the channel information should be designed to achieve the highest throughput over the connections between the cells.

The process of exchanging the traffic status between base stations is presented in Fig. 6-1.

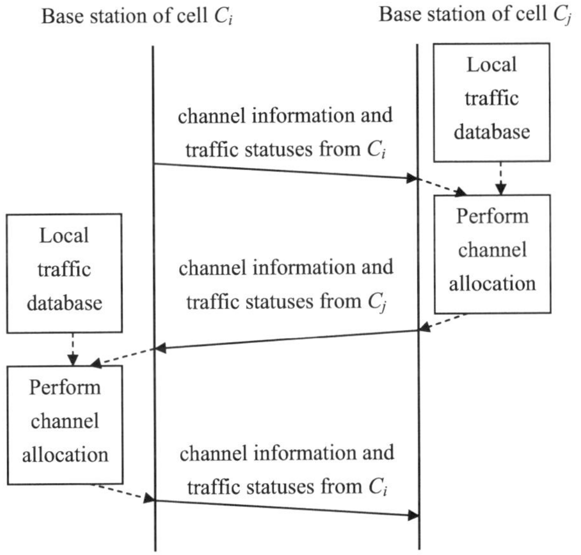


*Figure 6-1: Exchanging channel information and traffic status as well as the local channel allocation between two cells.*

#### 6.1.4.1 Transfer of the Resource States between Cells

For the channel allocation process, each base station (BS) of a cell is in one of the following 3 states: high number of channels per active user (*High-M*), low number of channels per active user (*Low-M*), or equilibrium (*Equil*). The BS state diagram including the conditions to change the states is presented in Fig. 6-2. When a BS has a higher $M$ compared to the lowest $M$ of its neighbors ($\min\{M_{BN}\}$), its state changes to *High-M* to calculate the number of channels to be released. After releasing channels, it state stays in *Equil*. If it has a lower $M$ compared to the lowest $M$ of its neighbor, its state changes to *Low-M* to seize the channels released by its neighbors. In the *Equil* state, if a BS detects free channels, it will additionally seize all these channels to gain transmission capacity of its slaves. If it detects interfering channels in use, the channels will be released. The state evaluation is trigged by the timer to send channel information (in a so-called channel announcement message, CA-Msg) or when the BS receives a new CA-Msg.

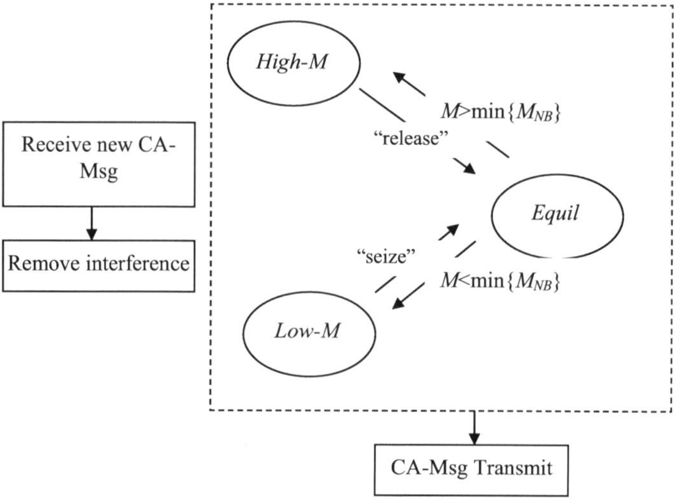


*Figure 6-2: BS transition state diagram*

### 6.1.4.2 Channel Information to be Exchanged

Each BS has to periodically broadcast its knowledge about the channels in a channel announcement message (CA-Msg). It is assumed that a time slot (one OFDM symbol) for transmitting the CA-Msg is organized immediately after the synchronization signal at the beginning of a time frame. The CA-Msg is followed by the channels in which the transmission of the PLC cells actually occurred. For a cell $C_i$, this message contains a group of fields; each data field provides the reception BS the following information:

- *AI*: 1 Byte for indicating if a cell is an access or inhome cell
- *ID*: 6 Bytes: address of the base station of the sending cell $C_i$
- $N_i$: 2 Bytes: number of active users of the sending cell $C_i$
- $H_i$: 1 Byte: number of neighbor cells of cell $C_i$
- $f_{List}^{(stt)}$: List of 2-bit-channel statuses, $f_{List}^{(stt)} = \{ f_k^{(stt)} : k \in [1, F] \}$. Value of $f_k^{(stt)}$ for a channel $f_k$:
    - 0: channel is not used
    - 1: occupied by this cell
    - 2: used by one neighbor cell
    - 3: used by more than one neighbor cell

The position and structure of a channel announcement message is presented in Fig. 6-3.

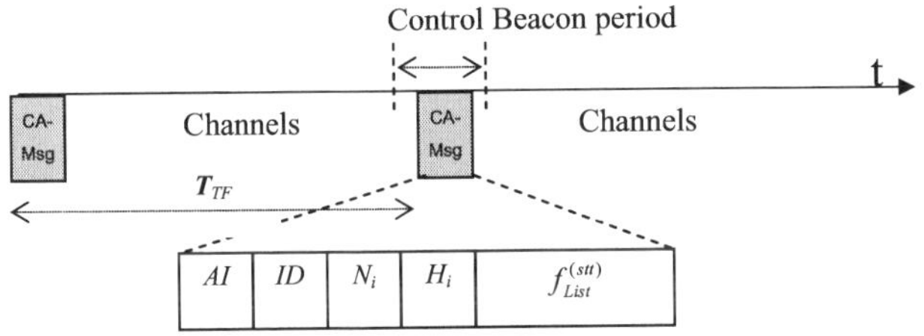


*Figure 6-3: Channel Announcement Message (CA-Msg) format*

Utilizing the status of the channels in the neighboring cells, the receiving cell can decide which channel to seize or release. Denote $f_k^{(stt)}(C_i)$ is channel status of channel $f_k$ received from cell $C_i$. In case a cell $C_j$ decides to seize a channel, it has to sort out and seize only the channel which is not being used by its neighbors ($f_k^{(stt)}$=0/ for all neighbor cells). If that cell $C_j$ decides to release a channel for its neighbor cell $C_i$, this channel is used by itself ($f_k^{(stt)}$=1) and should be used by only one neighbor cell $C_i$ ($f_k^{(stt)}(C_i)$=2). It means that this channel is used by only the neighbor cell $C_j$ of the cell $C_i$. When cell $C_j$ releases this channel $f_k$, it can be seized by cell $C_i$ without interfering with the other neighboring cells of cell $C_i$.

The length of a CA-Msg is calculated by the sum of all fields presented above. If $F$ denotes the number of available channels, $\lceil \cdot \rceil$ being the ceiling function, a CA-Msg requires

$$L = 1 + 6 + 2 + 1 + \left\lceil \frac{F \cdot 2}{8} \right\rceil = 10 + \left\lceil \frac{F}{4} \right\rceil \text{(Byte)} \qquad (6\text{-}16)$$

to transmit all of the necessary information. If there are $F$=80 channels for sharing between cells then, in accordance with (6-16), the length of a CA-Msg is $L$=30 bytes.

### 6.1.5 Processing of the Channel Information

#### 6.1.5.1 Procedures of the Channel Allocation Protocol

A diagram describing the processing of the CA-Msg inside a base station (BS) is presented in Fig. 6-4.

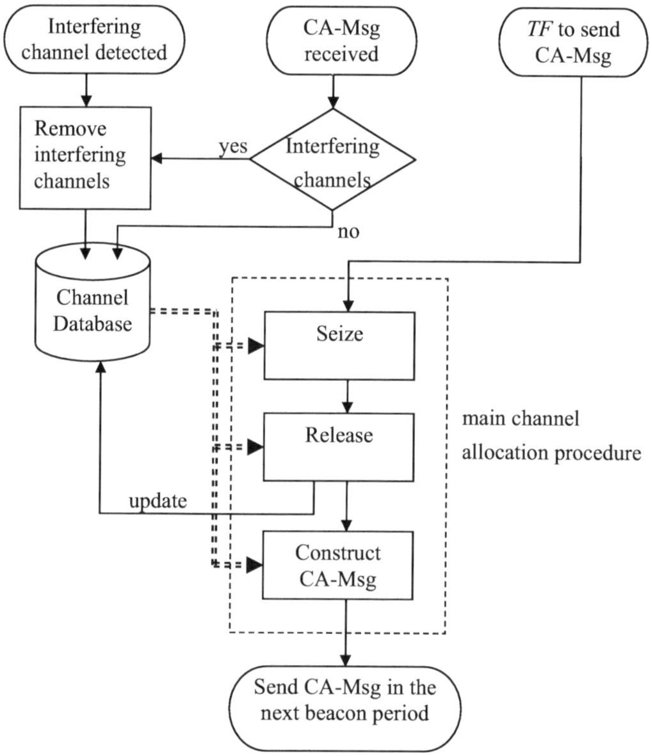


*Figure 6-4: Processing of the CA-Msg inside a base station*

When a CA-Msg is received, the BS has to check and immediately remove the interfering channels. The received channel information is updated in the channel database. The channel database is a table, which stores the entire channel information of its neighbors received via CA-Msgs and channel information of its cell. Each record in the table represents a neighbor cell and will be updated when a CA-Msg from the corresponding neighbor cell is received. Each record of a cell $C_j$ (identified by an $ID_j$) also contains the time of receiving the last CA-Msg from that cell, called $T_{j \to i}^{CA\_Msg}$, which allows the record to be removed when the message is outdated.

At a specific point in time, if a BS decides to send a CA-Msg in the incoming beacon time slot, it activates the main channel allocation procedure. Incorporated in this main procedure, the "Seize" block will examine whether this BS can obtain more channels for its slaves, the "Release" block will check whether this BS has to release channels to its neighbors, and the "Construct CA Message" block constructs the CA-Msg to be sent in the next beacon period.

| | | | | |
|---|---|---|---|---|
| $ID_x$ | $N_x$ | $H_x$ | $f_{List}^{(stt)}(C_x)$ | $T_{x \to i}^{CA\_Msg}$ |
| $ID_y$ | $N_y$ | $H_y$ | $f_{List}^{(stt)}(C_y)$ | $T_{y \to i}^{CA\_Msg}$ |
| … | | | | |

*Table 6-2: Channel Database*

#### 6.1.5.2 The "Seize" and "Release" Procedures

The main procedure of the protocol is based on the channel "Seize" and "Release" procedures.

a) *The channel "Seize" procedure.*

If the base station (BS) of a PLC cell requires additional channels for serving its users, the BS tries to seize more channels from the free channels in the network. A free channel from the perspective of a cell $C_i$ is a channel not used by any of its neighboring cells. A cell requests more channels when the allocated channels are not sufficient for the current real-time traffic requests or there is at least one elastic traffic connection, which requests as high transmission bandwidth as possible. By combining the lists of channels in use by its neighbor cells, the BS of the cell $C_i$ will create a list of free channels that it can seize for its internal transmission.

$$\text{search for channel } f_k \;/\; f_k^{(stt)}(C_j) = 0, \forall C_j \in \boldsymbol{C}_{nb}^{(i)} \qquad (6\text{-}17)$$

where $\boldsymbol{C}_{nb}^{(i)}$ is the set of neighbor cells of $C_i$.

The BS seizes all the free channels in order to maximize its number of reserved channels, which means also maximize the channel-reuse factor $f_{CR}$. The flow chart of the seize procedure is presented in Fig. 6-5a.

b) *The channel "Release" procedure.*

There are three situations in which a cell releases a channel: 1) interference is detected, 2) the cell has a higher $M$ value than its neighbors, or 3) it has at least one channel that is not in use. In the reference PLC system, the interference can be detected by a specific procedure, e.g. by means of the channel measurement procedure as described in chapter 2 and [Opera06]. The channels that may cause transmission interference with neighboring cells (used by at least one neighbor cell) will be released immediately. By analyzing the channel per active user received from its neighbor cells, the BS decides if it has to release one or several channels. The number channels to be released is calculated based on its number of channels per active user, compared to its neighbors.

With the objective to maintain fairness, the number of channels per user should be equal for all cells. The released channels are the channels not in use by the neighbors of its neighbor cells. Therefore, the released channels allow the neighbor cells to seize them without causing possible channel interference. A channel is also to be released if it is not in use by the BS, because the users in the cell currently have a sufficient number of channels for serving its current services, e.g. in case the cell has only a certain number of real-time traffic. A channel is also not useable when the transmission capacity of the channel in that cell is too low. The flow chart of the release procedure is presented in Fig. 6-5b.

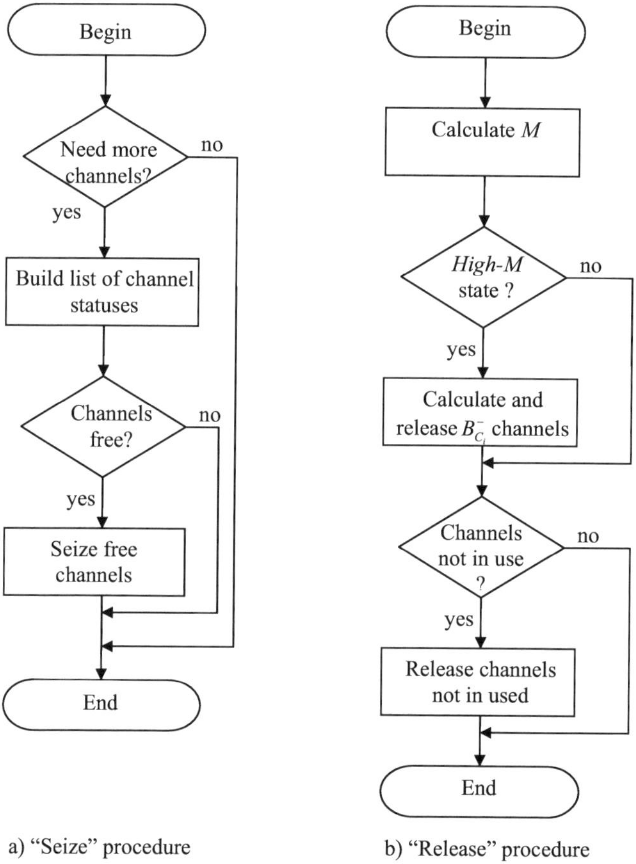


*Figure 6-5: "Seize" and "Release" procedures in a BS*

To find which cell has the lowest channels per user (cell $C_j$) by a BS, the comparison of $M_i$ is formulated as

$$\text{find } C_j \,/\, M_j = \min\{M_i \,/\, i = 1..S, i \neq j, i_{i,j} = 1\} \qquad (6\text{-}18)$$

If the BS finds that its cell is in the *Equil* or *Low-M* state, the release procedure releases a channel, only when this channel is not in use. Otherwise, in the *High-M* state, the BS of the cell $C_i$ observes that $M_i > M_j$ and has to determine the number of channels to be released. Let $B_{C_i}^-$ be the number of channels to be released by $C_i$, $B_{C_i}^-$ is also the number of channels that will be seized by the cell $C_j$. After the number of channels are released by $C_i$ ($M_i^{after\_release}$) and seized by the $C_i$ ($M_j^{after\_seize}$), both cells should have the same $M$ value. Therefore, $B_{C_i}^-$ should satisfy the following equation

$$M_i^{after\_release} = M_j^{after\_seize} \rightarrow \frac{B_{C_i} - B_{C_i}^-}{N_i} = \frac{B_{C_j} + B_{C_i}^-}{N_j} \tag{6-19}$$

Thus

$$B_{C_i}^- = \frac{B_{C_i} N_j - B_{C_j} N_i}{N_i + N_j} \tag{6-20}$$

However, the number of channels should be an integer, therefore, $B_{C_i}^-$ is rounded down to find the actual number of channels to be released to produce a relatively equal amount of channels per active user after the cell $C_j$ seizes these channels:

$$B_{C_i}^- = \left\lfloor \frac{B_{C_i} N_j - B_{C_j} N_i}{N_i + N_j} \right\rfloor \tag{6-21}$$

where $\lfloor \cdot \rfloor$ is the floor function.

Channels ($f_k$) to be released for $C_j$ when $B_{C_i}^- > 0$ are the channels not used by the other neighbors of $C_j$ ($f_k^{(stt)}(C_j)$=2). This information is received by $C_i$ in the channel announcement message sent by $C_j$. By sorting the average transmission capacities, the channel with the lowest transmission capacity is released first.

The pseudo-code showing the algorithm for making decision on allocated channels based on the "Seize" and "Release" procedures is summarized in Fig. 6-6.

```
begin //this cell is C_i
    if (time for sending CA-Msg)
    //seize block
        if (need more channels)
                build the list of channel statuses
                seize free channels (released by neighbor cells)
        endif
    //release block
        find a neighbor cell C_j with lowest M_j
        if (M_j< M_i) //release channels
                calculate B^-_{C_k} by (6-21)
                release B^-_{C_k} channels
        endif
        if (one or more channels not in use)
                release not in use channels
        endif
        construct and send out the CA-Msg
    else
        remove interfering channels
    endif
end
```

*Figure 6-6: Algorithm for making decision on transmission resource*

## 6.2 Communication between Neighbor Cells

The information to be exchanged will be transmitted in a reserved area inside every time frame (*TF*). To do so, a dedicated time slot of one OFDM symbol is assigned to the time frame for the transmission of the CA-Msg between the PLC cells. Each time frame starts with a control beacon time slot consisting of the synchronization signal for synchronizing between cells, e.g. as described in [HomePa07]. This is followed by an OFDM symbol reserved for transmitting of the CA-Msg and $F$ channels for sharing between PLC cells (Fig. 6-7). If the BS is located far from the neighbor cells, the transmission can be realized by the border nodes (BN). A border node (BN) is the node, which can receive signals from one or more neighbor cells.

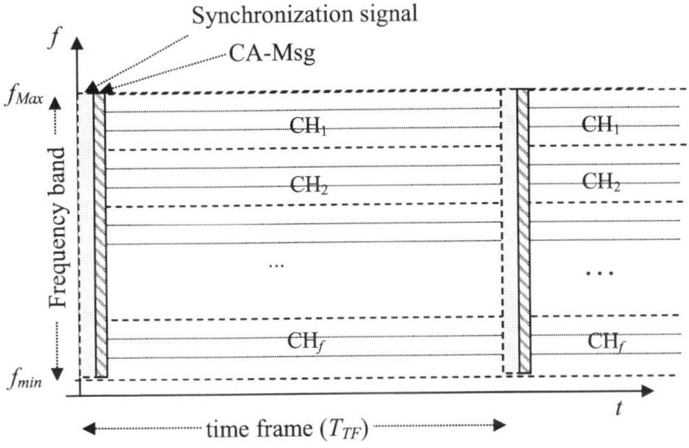


*Figure 6-7: Position of a CA-Msg in a time frame*

One consideration is to route the CA-Msg between the BSs; however, this requires overhead for the exchange of additional information between the nodes in both of the neighbor cells to specify the route. Since the network is dynamic, delivery of a CA-Msg may demand recovery of a new route. This may result in a long delivery time between the cells. Because of the complication in specifying a protocol for routing of the CA-Msg between the BSs of the neighbor cells, the CA-Msg is broadcasted from a specific BS in a selected time slot.

When the cells are synchronized with each other, the time slot for sending a CA-Msg is also synchronized for all the cells. When two neighbor cells choose the same beacon time slot (TS), two CA-Msgs collide and no CA-Msg is exchanged. Since both sending BSs do not detect the collision, they have to use the previously received channel information until the next CA-Msg successfully arrives or the previously channel information expires. The expiration of the received channel information is necessary in case the interfering node in the neighbor cell is turned off instead of CA-Msg collisions. Therefore, determining the transmission probability could reduce the collision probability ultimately affecting the transmission success. The decision to send a CA-Msg from a cell is presented in Sec. 6.2.4. Furthermore, the expiration of a CA-Msg has to be defined in order to avoid using invalid information.

The definition for transmission and detection of different transmission signals between the cells described in OPERA coexistence protocol [D28OP07] is used. Within the pre-defined common beacon time slot at the beginning of each time frame, the protocol organized the exchange of the channels information between the cells with the assumption that all cells are synchronized.

### 6.2.1 BS to BS Direct Communication

If the BSs from both neighbor cells, which are coupled into the same power line, are located near each other, direct transmissions between them may be realized. In this case, each BS detects the other and taking its neighbor's information for the information distribution. Each BS will select a beacon TS for transmitting it CA-Msg directly within the selected time frame (*TF*) (as presented in Fig. 6-8).

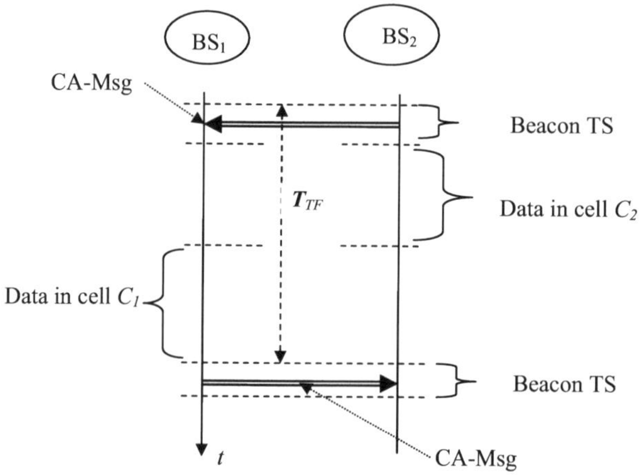


*Figure 6-8: BS to BS direct communication*

### 6.2.2 BS to BS Communication via Border Nodes

In case direct transmission between the BSs is not feasible, the border node (BN), which can detect the beacon signal from both the BSs will be involved in the CA-Msg transmission. One example showing the transmission of the CA message with the help of a user in the BS as the border node (BN) is presented in Fig. 6-9. The CA-Msg is transmitted from the $BS_1$ to the $BN_1$ during the normal downlink data transmission. The $BN_2$, after receiving the CA-Msg, will forward it in the next beacon period to the neighbor $BS_2$. If the BN receives the CA-Msg from the neighboring BS during the beacon time slot, it will include the CA-Msg in the next uplink data transmission to its BS.

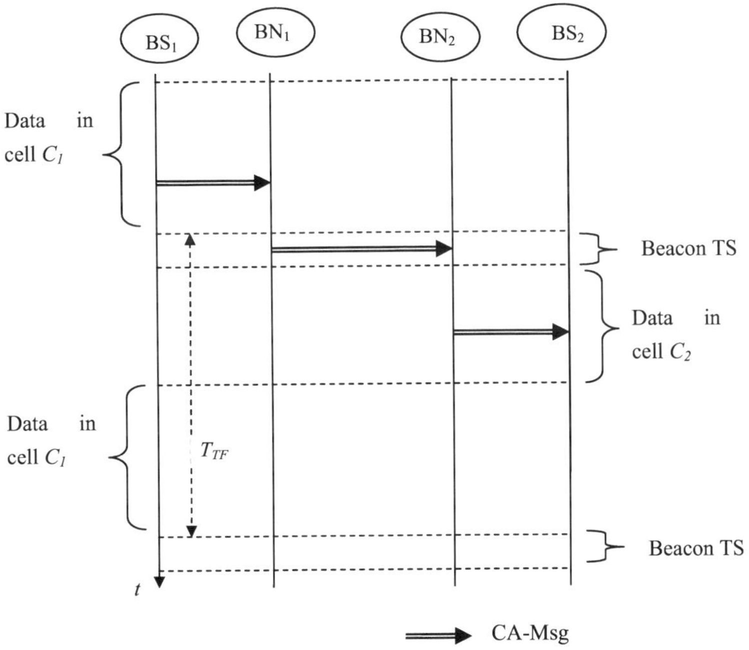


*Figure 6-9: BS to BS communication via border nodes*

### 6.2.3 Broadcasting of the CA-Msg from a Cell

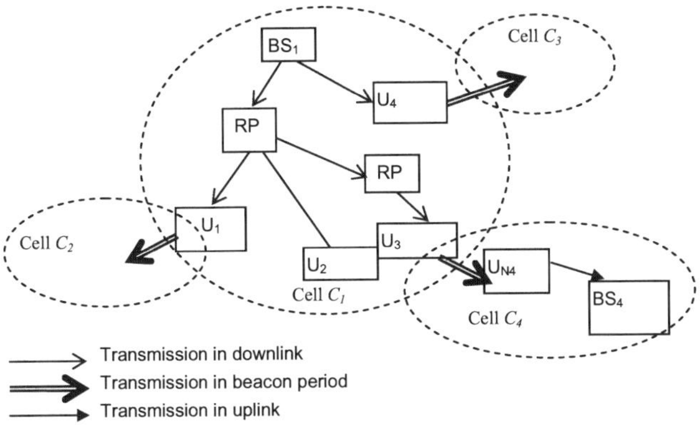


*Figure 6-10: Broadcast from elements in a cell*

As previously specified, a CA-Msg is sent out from a cell to its neighbors in a broadcasting fashion. Each BS of a cell receives interfering information from its border nodes to update its channel database. When a cell decides to send a CA-Msg in the next beacon TS, the BS builds the message and sends it to all the users within the normal downlink transmission. These users, when receiving a CA-Msg from its master, send the same CA-Msg in the incoming beacon TS. When receiving this message in the neighbor cell, its border node will send the message to its master in the normal uplink transmission. One example of the transmission of a CA-Msg from the BS in the Cell $C_1$ to three neighbor cells is presented in Fig. 6-10. User $U_1$, $U_3$, and $U_4$ act as border nodes and send the same CA-Msg to the neighbor cells in the same beacon TS.

More than one PLC cell sending different CA messages in the same beacon TS may cause a collision. The time slot chosen by a BS for sending its CA-Msg has to be selected in order to achieve the highest reception rate (the number of received CA-Msg per time frame) by its neighbors.

### 6.2.4 Decision to Send CA-Msg

The channel announcement message (CA-Msg) is sent randomly from a cell $C_i$ in a beacon time slot (time slot for short) with a probability $p_i^{send}$, similar to the principle of the slotted Aloha protocol [Chan00]. Since the BS can not monitor whether or not a neighbor receives a message after sending a message, there is no mechanism for retransmission. If there are $H_i$ neighbors of a cell $C_i$, there are $H_i$ +1 interfering cells from the point of view of the $C_i$ (Fig. 6-11). Assuming that all $H_i$ +1 neighbor cells send with the same probability $p_j^{send}$, the delivering message from $C_j$ to $C_i$ is only successful if $C_j$ sends while the other $H_i$ cells do not send. Therefore, the probability of a successful packet transmission from $C_j$ to $C_i$ is

$$p_{j\to i}^{recv} = p_j^{send}\left(1 - p_j^{send}\right)^{H_i} \tag{6-22}$$

By solving (6-22) to find Max $\{p_{j\to i}^{recv}\}$, the probability that each neighbor $C_j$ chooses to send the message which achieves the highest success reception rate as calculated in (6-22) at the $C_i$ is

$$p_j^{send} = \frac{1}{H_i + 1} \tag{6-23}$$

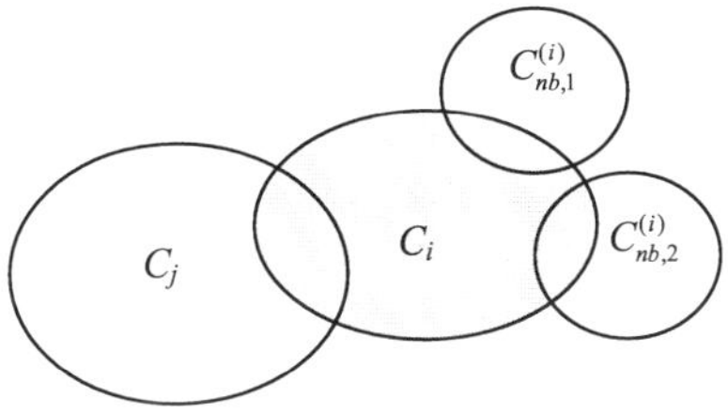


*Figure 6-11: Cell $C_i$ with $H_i$=3*

The average duration for the delivery of a message from $C_j$ to $C_i$ is

$$D_{j\to i} = 1/p_{j\to i}^{recv} \tag{6-24}$$

Based on this calculation, when the success reception rate at $C_i$ is at its highest value, the delay of delivering a message in that connection is the shortest. For three examples of a cell $C_i$ which has 1, 2, and 8 neighbors respectively, the reception rate at $C_i$ is presented in Fig. 6-12a, the average packet delivery time is presented in Fig. 6-12b.

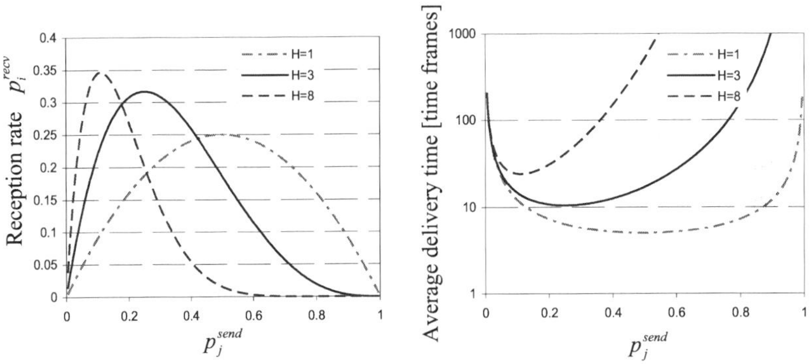


*Figure 6-12: a) Reception rate $p_i^{recv}$, b) Average delivery time $\overline{D_{Ch}}$ : dependency with $p_j^{send}$ by neighbors of $C_i$*

In a PLC network, a cell $C_j$ may have different number of neighbor cells, thus it has to choose a particular $p_j^{send}$ based on its neighbors' situations. Denote $\boldsymbol{C}_{nb}^{(i)}$ is the set of neighbors of cell $C_i$, each neighboring cell is indexed from 1 to $H_i$, the probability of successful packet transmission from $C_j$ to $C_i$ is

$$p_{j\to i}^{recv} = p_j^{send}(1-p_i^{send}) \prod_{k=1,k\neq j}^{H_i} \left(1-p_k^{send}\right) \tag{6-25}$$

If the reception rate at a cell $C_i$ is defined as the average number of messages received per time frame, then it is calculated by the sum of the reception probabilities in a time frame from all neighbors

$$p_i^{recv} = \sum_{j=1}^{H_i} p_{j\to i}^{recv} = \sum_{j=1}^{H_i}\left( p_j^{send}(1-p_i^{send}) \prod_{k=1,k\neq j}^{H_i} \left(1-p_k^{send}\right)\right) \tag{6-26}$$

It is also defined as the throughput of the CA-Msg at the cell $C_i$. The average reception rate of all the cells in the network is

$$\overline{p^{recv}} = \frac{1}{S}\sum_{i=1}^{S} p_i^{recv} = \frac{1}{S}\sum_{i=1}^{S}\sum_{j=1}^{H_i} p_j^{send}(1-p_i^{send}) \prod_{k=1;k\neq j}^{H_i} \left(1-p_k^{send}\right) \tag{6-27}$$

Using (6-15) and (6-24), the average delivery time in all the links between each pair of neighboring cells is

$$\overline{D_{Ch}} = \frac{1}{n^{conn}}\sum_{i=1}^{S-1}\sum_{j=i+1}^{S} i_{i,j}\frac{1/p_{j\to i}^{recv} + 1/p_{i\to j}^{recv}}{2} \tag{6-28}$$

Considering that $H_i = \sum_{j=1;j\neq i}^{S} i_{i,j}$, (6-28) can be rewritten as

$$\overline{D_{Ch}} = \frac{1}{2n^{conn}}\sum_{i=1}^{S}\sum_{j=1}^{H_i}\left(1/p_{j\to i}^{recv}\right) = \frac{1}{2n^{conn}}\sum_{i=1}^{S}\sum_{j=1}^{H_i}\frac{1}{p_j^{send}(1-p_i^{send})\prod_{k=1;k\neq j}^{H_i}\left(1-p_k^{send}\right)} \tag{6-29}$$

where $n^{conn}$ , as calculated in (6-14), is the number of end-to-end connections between two BSs of two cells.

In case cell $C_j$ has more than one neighbor, it has to select an appropriate $p_j^{send}$ in order to achieve the minimum average delivery time to all of its neighbors. Considering the example network presented in Fig. 6-11 and the results in Fig. 6-12, if cell $C_j$ has three neighbors, each neighbor also has its own number of neighbors, then one selected $p_j^{send}$ does not minimize the average delivery time of all of its neighbors. If it aims to minimize the average delivery time of the CA-Msg to a neighbor which only has one neighbor ($H$=1), then $p_j^{send}$ is 1/($H$+1) = 1/2 = 0.5; to minimize the average delivery time to a neighbor with $H$=3, then $p_j^{send}$ is 1/($H$+1) = 1/4 = 0.25; and for a neighbor with $H$=8, $p_j^{send}$ is 1/($H$+1) = 1/9 $=0.\overline{1}$. In accordance with (6-28), it is too complex to calculate $p_j^{send}$ of each BS in a cell in order to achieve the shortest average delivery time $\overline{D_{Ch}}$ for the whole network. Therefore, two simple methods are considered to calculate the $p_j^{send}$, that provide an adequate shortest average delivery time to a cell $C_i$ ($\overline{D_{\rightarrow i}}$) locally.

- Using the Minimum Sending Probabilities

By experimenting with the numerical calculation for equation (6-29) with different value of $H_i$ and $H_j$, one of the simple calculations which returns a short $\overline{D_{\rightarrow i}}$ is to use the minimum of the $p_j^{send}$:

$$p_j^{send} = \min_{C_i \in C_{nb}^{(j)}} \left( \frac{1}{H_i + 1} \right) \tag{6-30}$$

- Using the Average Sending Probabilities

Because a cell $C_j$ has to balance the successful reception rates of the channel announcement messages with all of its neighbors, the average value of the sending probabilities is selected. Therefore, the probability to send from the cell $C_j$ is

$$p_j^{send} = \frac{1}{H_j} \cdot \sum_{C_i \in C_{nb}^{(j)}} \frac{1}{H_i + 1} \tag{6-31}$$

When the BS of the cell $C_j$ receives the channel announcement message, it obtains the necessary inputs from its neighbors to calculate the $p_j^{send}$ in (6-31). If there is no neighboring network, the cell sends its CA-Msg with $p_j^{send} = 0.5$.

Both methods used for calculating the sending probability have advantages in specific cases. The PLC network reference models however have a shorter average delivery delay when applying (6-31). Therefore, this equation is used in the evaluation parts.

### 6.2.5 Lifetime of the CA-Msg

In order to keep the channel information received from a neighbor cell up-to-date, it is necessary to setup a timer for the CA-Msg received from a neighbor cell. When the timer expires, the information of the corresponding cell will be removed from the channel database. If the lifetime is too long, the information concerning this neighbor without any updates over a long period of time may be invalid. If it is too short, the neighbor's information may be deleted when the neighbor cell is still there and the channel announcement message is lost due to transmission collision.

By monitoring the message reception, the BS can calculate the average delay of a successful transmission based on the slotted Aloha protocol. The unit for the validity is time frame $T_{TF}$. The theoretical delay caused by send slot selection and the average number of times sent per successful transmission, under the maximum throughput condition is derived from [Chan00]:

$(e-1)/p_i^{send} \cdot T_{TF}$, in which $p_i^{send}$ is the sending probability of the CA-Msg from cell $C_i$ to achieve maximum CA-Msg throughput of its neighbor. Therefore, lifetime of the timer for a message arriving from cell $C_i$ is defined as equal to two times that duration

$$T_i^{ex} = 2\frac{e-1}{p_i^{send}} \cdot T_{TF} \tag{6-32}$$

### 6.2.6 Bandwidth for the CA-Msg Communication

We calculate the required bandwidth as the necessary overhead when introducing the channel announcement message (CA-Msg) in every time frame $T_{TF}$. In accordance with the transmission frame definition in the OPERA specification [Opera06] and the coexistence frame defined in [D18OP05] for the 50 Hz electricity power supply network, the duration of a frame $T_{TF}$ =240 ms is chosen. For $F$=80 channels, the length of the channel announcement packet is $L$=30 bytes, in accordance with the calculation in section 6.1.4.2. With the binary phase shift keying (BPSK) modulation scheme, the packet can be transmitted in one OFDM symbol with the duration of 71.2 µs [Opera06]. Therefore, the ratio of the overhead compared to a time frame is 71.2µs/240 ms $\approx 0.3 \cdot 10^{-3}$. This ratio shows that the overhead generated by introducing the channel announcement message is very small and thus it is negligible.

## 6.3 Performance Evaluation

### 6.3.1 Simulation Model

Due to the model's complexity, a simulative performance analysis has been selected. The proposed negotiation strategy is built into the YATS simulator [Baum08]. YATS is a discrete event, discrete time simulation tool. It is used for investigation of various communication networks. It was developed at the Chair for Telecommunications, Technische Universität Dresden. In the area of PLC, the simulator is used to simulate several communication protocols. It has been used by the author, along with other authors, to investigate and evaluate the performance of different multiple access control (MAC) protocol for narrowband PLC networks [DoHr05] [DoHr06] [DoHr07] as well as for broadband PLC networks in the OPERA2 project [D22OP08].

#### 6.3.1.1 Network Model

For analyzing the performance parameters and the change in users' statuses, the reference PLC networks in Sec. 4.2.1 are taken into account. Fig. 6-13 presents the generic simulation model. A simulation network model contains a PHY network layer, several base stations connected to that layer, and several users connected to each base station. The CA-Msgs are transmitted between the BSs through the PHY network layer. The PLC network layer is represented by a PER connection matrix, as presented in section 4.4.1, and works based on the given PER matrix of a specific network scenario. All the BSs are synchronized and run the proposed channel allocation protocol.

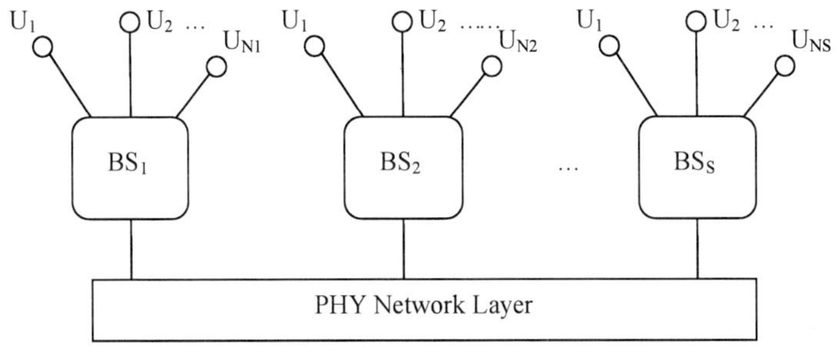


*Figure 6-13: Simulation model for multi-cell PLC network*

#### 6.3.1.2 User Model

At each moment, a user can be in one of two states in accordance with the ON-OFF model as presented in Fig. 6-14:

- Active: modeled by the ON state. This user requests transmission capacity for its transmission.
- Inactive: modeled by the OFF state. This user is in idle status or turned-off. It does not require transmission capacity at this moment.

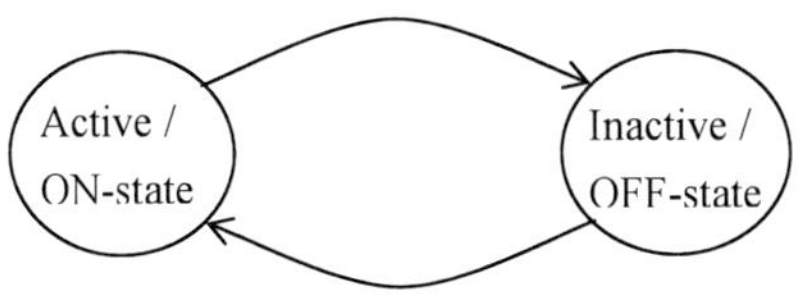


*Figure 6-14: Active and inactive states in accordance with the ON-OFF model*

Generally, the duration of the ON and OFF phases depends on the duration of the data transmission from user's services. As previously presented in Sec. 4.4.2, since the data arriving in a discrete time frame, the ON and OFF periods of a user are modeled by the geometric distribution. The mean values of the durations are given by a number of time frames.

#### 6.3.1.3 Simulation Parameters

If not explicitly mentioned, each simulation result is conducted from 10 simulation sub-runs, each lasting 100.000 time frames. This means that each subrun simulates a PLC network for 100.000×240 ms=400 minutes. Mean values are calculated. The statistical accuracy is calculated with a confidence probability of 0.95. Each cell contains 20 users, and the status of a user is assumed to alternate between active and inactive (modeled in Sec. 6.3.1.2) with the mean duration of 800 time frames (800×240 ms=3.2 minutes). This means that in each subrun, a user changes its status approximately 400/3.2=125 times.

### 6.3.2 Channel Re-Allocation between PLC Cells

The re-allocation of the channels when applying the proposed DICRA protocol is demonstrated and analyzed by means of a sample network which contains only two neighboring cells. For this purpose, cell $C_1$ contains 20 users, and the number of active users in this cell increases and decreases consequently between 1 and 20. The change of the number of active users occurs every 100 time frames ($T_{TF}$). Cell $C_2$ has 10 active users during the run-time. During the simulation, each cell detects one neighbor. In accordance with (6-31), both cells send their CA-Msg with a probability in a time frame of 0.5 ( $p_1^{send} = p_2^{send}$ =0.5).

#### 6.3.2.1 Number of Allocated Channels

The time series of the number of reserved channels in cell $C_1$ and cell $C_2$ during the first 8,000$T_{TF}$ are presented in Fig. 6-15a and Fig. 6-15b. The simulation results show that the number of channels in both cells adapts to the variation of the number of users of cell $C_1$. In cell $C_1$, the change in the number of the reserved channels ($B_1$) displays the same trend of the variation of the number of active users ($N_1$) while in cell $C_2$, the number of reserved channels ($B_2$) varies without the change in the number of active users ($N_2$). As expected from the protocol, we observe a dependency of the number of reserved channels in cell $C_2$ with the number of active users in cell $C_1$: the number of reserved channels $B_2$ increases when the number of user $N_1$ decreases and $B_2$ decreases when the number of $N_1$ increases.

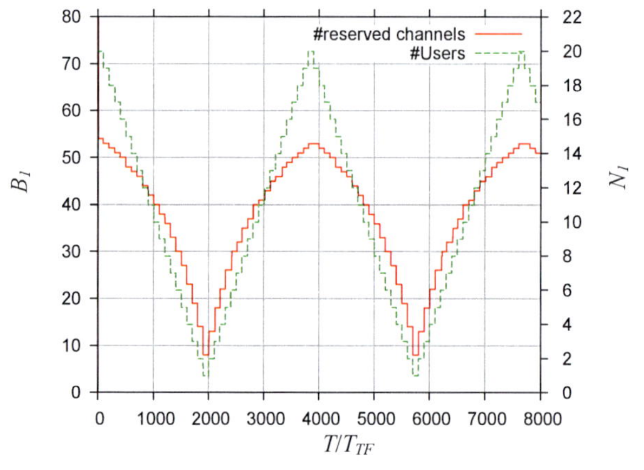


a)

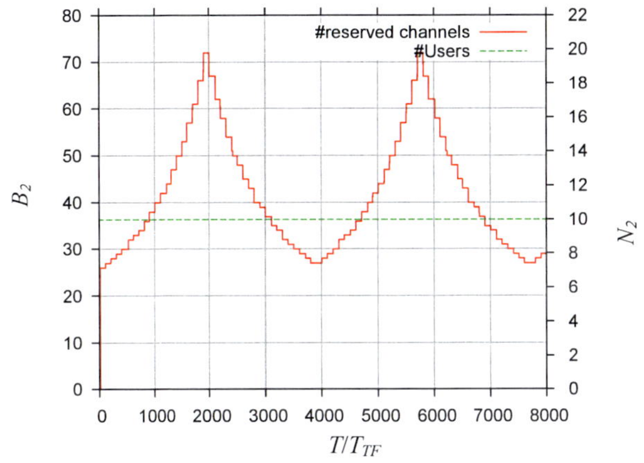


b)

*Figure 6-15: Variations of the number of reserved/allocated channels; a) in cell $C_1$; b) in* cell $C_2$

### 6.3.2.2 Analysis of the Variation of the M Values

The cause for the variation of $B_2$ to the change of $N_1$ is the change in the cells' states based on the comparison of the $M$ values between the cells. Fig. 6-16a and Fig. 6-16b present the variations of the number of channels per active user ($M$ values) in cell $C_1$ and cell $C_2$ during the first 8,000 $T_{TF}$. During run-time, the protocol aims to keep the same $M$ values in the both cells.

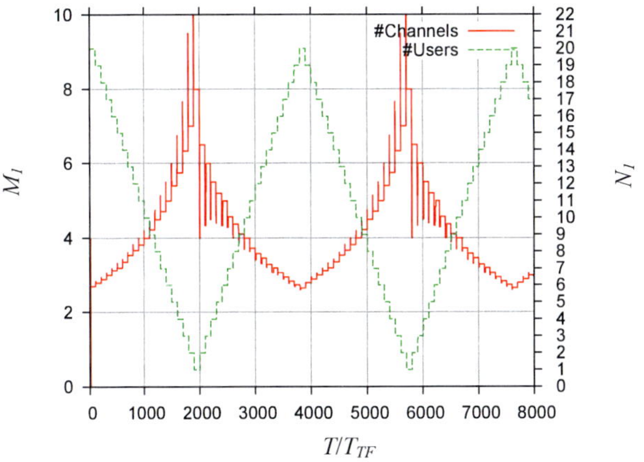


a)

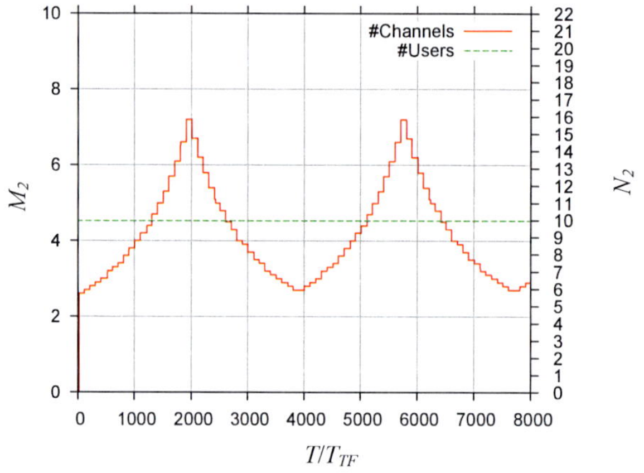


b)

*Figure 6-16: Variations of the number of channels per active user; a) in cell $C_1$; b) in* cell $C_2$

There are two phases, an increasing- and a decreasing phase, of the number of active users $N_1$. Fig. 6-17 presents a closer look over a period of $1000T_{TF}$ at the $M$ values in both cells in these two phases. When the number of active users ($N_1$) decreases, the $M_1$ also increases (e.g. point $X_1$ in Fig. 6-17a). Because this value is higher than the $M_2$ of cell $C_2$ (*High-M* state), cell $C_1$ releases some channels, e.g. at the point $X_2$ in Fig. 6-17a, for cell $C_2$. The number of channels to be released is calculated in accordance with (6-21). After releasing channels, $M_1$ reduces but it is still higher than the value in the period where $N_1$ is higher (point $X_0$ in Fig. 6-17a). The value of $M_2$ of the cell $C_2$ also increases (point $X_3$ in Fig. 6-17c) when cell $C_2$ seizes the channels that cell $C_1$ has just released. At the end of the adjustment period, both cells have the same $M$ and they are in the *Equil* state.

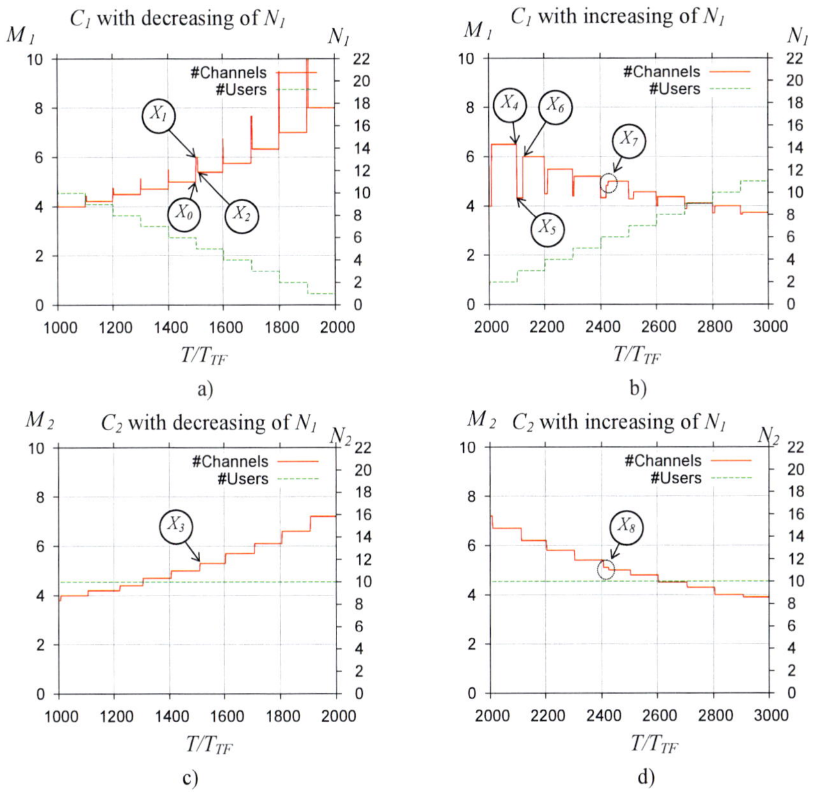


*Figure 6-17: Channel per user variations; a) cell $C_1$ with decreasing of $N_1$; b) cell $C_1$ with increasing of $N_1$; c) cell $C_2$ with decreasing of $N_1$; d) cell $C_2$ with increasing of $N_1$*

In the increasing phase, $N_1$ increases causing $M_1$ to decrease, e.g. from the point $X_4$ to the point $X_4$ in Fig. 6-17b. When the CA-Msg from cell $C_1$ is sent to cell $C_2$, cell $C_2$ observes that it is in *High-M* state and thus releases some channels corresponding with (6-21), and sends its CA-Msg to cell $C_1$. When cell $C_1$ receives the channel information from cell $C_2$, it is in the *Low-M* state and thus seizes the channels that have been previously released by cell $C_1$, and its $M_1$ increases to the value in the point $X_6$ in Fig. 6-17b. In the results, with the increasing of $N_1$, both $M_1$ and $M_2$ reduce in both of the cells after the CA-Msgs have been exchanged, and both cells are in the *Equil* state.

By applying the floor function in (6-21) to calculate the number of channels to be released, in some circumstances, two cells have to perform the release and seize procedures two times to be in the *Equil* state. This situation is observed by the variation of the $M$ values in an example with two cycles pointed out by $X_7$ in Fig. 6-17b and $X_8$ in Fig. 6-17d.

### 6.3.2.3 Analysis of the Adaptation Time

The time required to vary the number of channels to adapt to the change of the number of active users in a cell is called the adaptation time. In this analysis, the adaptation times in cell $C_1$ and cell $C_2$ in two cases with a varying number of active users $N_1$ are presented in Fig. 6-18.

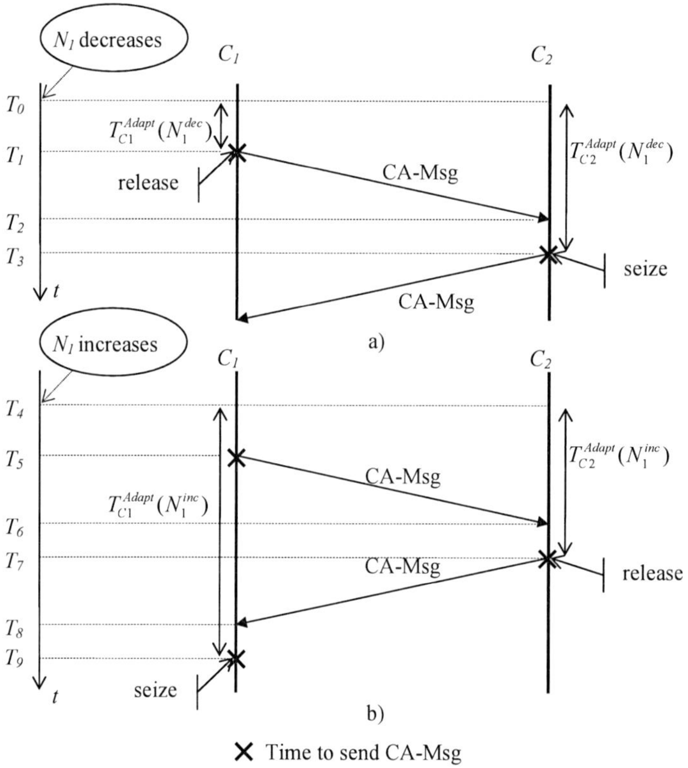


*Figure 6-18: Adaptation of channels to the variation of user; a) with decreasing of $N_1$; b) with increasing of $N_1$*

In case $N_1$ decreases, e.g. at the point of time $T_0$ in Fig. 6-18a, according to the protocol (Sec. 6.1.4) cell $C_1$ releases one or some channels in the time $T_1$ before sending out its CA-Msg. The duration for the number of channels in $C_1$ to adapt to the change of $N_1$ is $T_{C1}^{Adapt}(N_1^{dec}) = T_1 - T_0$. Assume that the CA-Msg is received by the BS of the cell $C_2$ at $T_2$ and the free channels are seized in a time frame $T_3$, the adaptation time of cell $C_2$ to the change of

$N_1$ is $T_{C2}^{Adapt}(N_1^{dec}) = T_3 - T_0$. This value is equal to the adaptation time of the cell $C_1$ plus the transmission time of the CA-Msg and the waiting time to send out a CA-Msg.

The adaptation process for the case in which $N_1$ increases is presented in Fig. 6-18b. Similar to the case above, if $N_1$ increases at $T_4$, the CA-Msg informing of the new number of active users from $C_1$ is sent to $C_2$ at $T_5$, and is received by $C_2$ at $T_6$. When the cell $C_2$ decides to send its CA-Msg at $T_7$, it also releases channels at this point of time. The adaptation time for changing the number of channels in $C_2$ to the variation of $N_1$ is $T_{C2}^{Adapt}(N_1^{inc}) = T_9 - T_4$. Subsequently, at $T_8$, the message from $C_2$ is received by the cell $C_1$, and $C_1$ seizes the free channels that have been released by $C_2$ at $T_9$. Therefore, the adaptation time for changing the number of channels in $C_1$ to the variation of $N_1$ in this case is $T_{C1}^{Adapt}(N_1^{inc}) = T_9 - T_4$.

By running a simulation for the change of the number of active users $N_1$ every 100 *TF* in this network with 10 sub-runs, where each sub-runs is 20,000 *TF*, so $N_1$ increases or decreases in a sub-runs 20,000/100/2=100 times. These adaptation times are measured. The simulation results are presented in Tab. 6-3.

| *Measurement Value:* | *Mean/$T_{TF}$* | *Confidence Interval /$T_{TF}$* |
|---|---|---|
| $T_{C1}^{Adapt}(N_1^{inc})$ | 10.171 | [9.853, 10.488] |
| $T_{C2}^{Adapt}(N_1^{inc})$ | 6.260 | [5.972, 6.549] |
| $T_{C1}^{Adapt}(N_1^{dec})$ | 2.290 | [2.120, 2.460] |
| $T_{C2}^{Adapt}(N_1^{dec})$ | 6.239 | [5.825, 6.654] |

*Table 6-3: Adaptation time in a network with 2 cells*

The largest portions of the adaptation time are the transmission times between the cells, including the waiting time until the time frame for sending the CA-Msg. The average adaptation time for each cell is also calculated, for the cell $C_1$: $T_{C1}^{Adapt}$ =6.450 *TF* and for the cell $C_2$: $T_{C2}^{Adapt}$ =6.250 *TF*. This means that for this network, when there is a change in the number of users in a cell $C_i$, it requires an average of 6.450 *TF* = 6.450×240 ms≈1.55 s to adapt to the number of allocated channels while the average adaptation time of its neighbor is 6.250 *TF* = 1.50 s.

### 6.3.2.4 Utilization of the Channel Information in CA-Msg

For analyzing the efficiency of the "Seize" and "Release" procedures, a PLC network consisting of 3 cells is selected, as presented in Fig. 6-19. The topographically neighboring cells interfere with each other (cell $C_1$ and cell $C_2$, cell $C_2$ and cell $C_3$). Cell $C_1$ and cell $C_3$ are far enough from each other and thus do not interfere with each other and can use the same channels at the same time. In this network, if $C_1$ releases a channel for $C_2$ which is currently being used by $C_3$, cell $C_2$ cannot seize it. This network allows for analysis of the case in which a cell has to find channels to release providing that for its neighbor to seize. For example, in this network, $C_1$ has to find channels which are not used by $C_3$ to release for $C_2$.

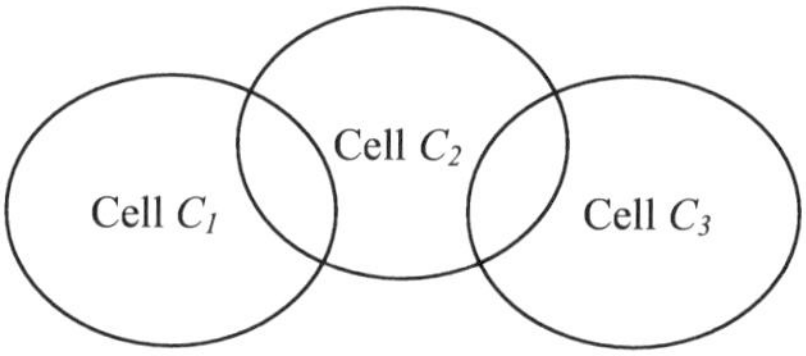


*Figure 6-19: Network in chain structure with 3 cells*

For describing the network connectivity, the interference matrix of the network is

$$\boldsymbol{I}_{in}=\begin{bmatrix} 0 & 1 & 0 \\ 1 & 0 & 1 \\ 0 & 1 & 0 \end{bmatrix} \tag{6-33}$$

Assume that when there is no activity in the network, two neighbor cells can exchange CA-Msg without error, so the packet error ration between two neighbor cells is

$$PER(C_i,C_j)=\begin{cases} 0 & \text{if } i_{i,j}=1 \\ 1 & \text{otherwise} \end{cases} \tag{6-34}$$

Each cell $C_i$ selects the beacon time slot for sending its CA-Msg message, based on the probability ($p_i^{send}$) of selecting a time slot by using (6-31). Cell $C_1$ and cell $C_3$ only have one neighbor, thus $H_1$=$H_3$=1, cell $C_2$ has 2 neighbors, therefore $H_2$=2. Based on received CA-Msgs, each cell calculate its $p_i^{send}$

In cell $C_1$: $p_1^{send}=\frac{1}{H_1}\cdot\frac{1}{H_2+1}=0.\overline{3}$, in cell $C_2$: $p_2^{send}=\frac{1}{H_2}\cdot\left(\frac{1}{H_1+1}+\frac{1}{H_3+1}\right)=0.5$, and in cell $C_3$: $p_3^{send}=\frac{1}{H_3}\cdot\frac{1}{H_2+1}=0.\overline{3}$

One situation is selected to demonstrate of the functionalities of the "Seize" and "Release" procedures with the coding of the channels in the $f_{List}^{(stt)}$ in a CA-Msg. Assume that each cell has 6 users and there are 6 channels to be used by this network model.

- Assume that at one point in time, cell $C_1$ and $C_3$ utilize some channels as in the following table. If cell $C_i$ uses the channel $F_j$, the appropriate position in the table is represented by "1". Cell $C_1$ uses channels $f_1, f_2, f_3, f_5, f_6$. Cell $C_3$ uses the channels $f_2, f_5, f_6$.

| | $f_1$ | $f_2$ | $f_3$ | $f_4$ | $f_5$ | $f_6$ |
|---|---|---|---|---|---|---|
| Cell $C_1$ | 1 | 1 | 1 | 0 | 1 | 1 |
| Cell $C_3$ | 0 | 1 | 0 | 0 | 1 | 1 |

- When cell $C_2$ is activated, by receiving the CA-Msgs from the cell $C_1$ and cell $C_3$, it knows the channels are being used by its neighbors, marked as "x" in the following table. It can only use the last channel $f_4$ when its "Seize" procedure is executed. Subsequently, each channel is coded with one of 4 statuses: 0: channel is not used, 1 channel is used by this cell, 2 channels are used by its neighbor, and 3 channels are used by at least 2 neighbors. In this case, channels $f_1$ and $f_3$ are coded with status "2", as presented in following table:

| | $f_1$ | $f_2$ | $f_3$ | $f_4$ | $f_5$ | $f_6$ |
|---|---|---|---|---|---|---|
| Used by neighbors | x | x | x | | x | x |
| Cell $C_2$ | 2 | 3 | 2 | 1 | 3 | 3 |

- In the following moment, it is assumed that all cells have the same traffic request (also assumed that each cell has 6 active users). When cell $C_1$ receives a CA-Msg from cell $C_2$, it builds its channel database as presented in following table

| | $f_1$ | $f_2$ | $f_3$ | $f_4$ | $f_5$ | $f_6$ |
|---|---|---|---|---|---|---|
| Cell $C_1$ | 1 | 1 | 1 | 2 | 1 | 1 |
| Cell $C_2$ | 2 | 3 | 2 | 1 | 3 | 3 |

- When the "Release" procedure in cell $C_1$ is executed, it observes that it has more channels per users (5/6) than cell $C_2$ (1/6); therefore, it should release (using (6-21))

$$B_{C_1}^{-} = \left\lfloor \frac{B_{C_1}N_2 - B_{C_2}N_1}{N_1 + N_2} \right\rfloor = \left\lfloor \frac{5\times 6 - 1\times 6}{6+6} \right\rfloor = 2 \text{ (channels)}$$

- Furthermore, cell $C_1$ observes that of its channels in use, only $f_1$ and $f_3$ are not being used by any other neighbor of cell $C_2$ (represented by "2" in the CA-Msg from cell $C_2$); therefore, it releases the channels $f_1$ and $f_3$. After releasing these channels (marked with "0"), cell $C_1$ uses only $f_2$, $f_5$ and $f_6$.

| | $f_1$ | $f_2$ | $f_3$ | $f_4$ | $f_5$ | $f_6$ |
|---|---|---|---|---|---|---|
| Release | x | | x | | | |
| Cell $C_1$ | 0 | 1 | 0 | 2 | 1 | 1 |

- When cell $C_2$ receive the CA-Msg from cell $C_1$, it notices the free channels $f_1$ and $f_3$. These channels are not used by any of its neighbors; therefore, cell $C_1$ seizes these channels $f_1$ and $f_3$. After these steps for performing channel reallocation, each cell has 3 channels reserved for their 6 active users. The number of channels per user is the same for 3 cells, each active user is reserved 3/6 = 1/2 channels.

| | $f_1$ | $f_2$ | $f_3$ | $f_4$ | $f_5$ | $f_6$ |
|---|---|---|---|---|---|---|
| Cell $C_1$ | 2 | 1 | 2 | 2 | 1 | 1 |
| Cell $C_2$ | 1 | 3 | 1 | 1 | 3 | 3 |
| Cell $C_3$ | 2 | 1 | 2 | 2 | 1 | 1 |

### 6.3.3 Scenario 1: PLC Cells in a Chain Structure

#### 6.3.3.1 Throughput and Delay of the CA-Msg

For evaluation of the effectiveness of the exchange of channel information between the cells, the throughput and delay of the transmissions for the CA-Msg are observed by means of both simulation and analysis methods. The simulation with the PLC networks which consists of a different number of cells placed in the chain structure is performed. Each simulation result for a PLC network is shown by its mean and confidence interval.

The average throughput of the CA-Msg in each connection $\overline{p^{recv}}$ and the corresponding average delivery time $\overline{D_{Ch}}$ are also calculated by means of the equations (6-27) and (6-28).

The average delivery time and the average throughput are also analyzed for different number of cells, from 2 to 8 cells, in a chain structure. The simulation results and the calculated results are presented in the same plots for comparisons purpose, as presented in Fig. 6-20. The calculated results are within the confidence intervals of the corresponding simulation results.

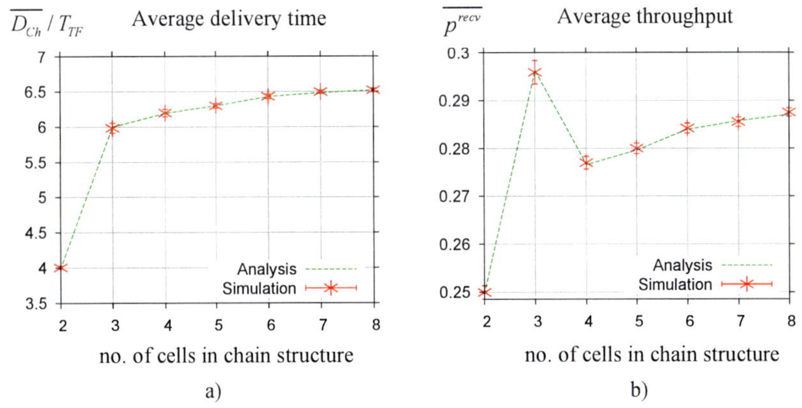


*Figure 6-20: a) Average delivery time; and b) average throughput of the CA-Msg in networks with a chain structure*

As presented in Fig. 6-20a, the average delivery delay of the CA-Msg increases when the number of the cells in the network increases. However, there are the limit values of $\overline{D_{Ch}}$ and $\overline{p^{recv}}$ when the number of cells $S$ in the network increases. These values are found by means of (6-27) and (6-28), and presented in Appendix A.2.1:

$$\lim_{S\to\infty}(\overline{D_{Ch}}) = 6.75 \text{ and} \tag{6-35}$$

$$\lim_{S\to\infty}(\overline{p^{recv}}) = 8/27 = 0.\overline{296} \tag{6-36}$$

There are large variations of the sending probability of the CA-Msg ($p^{send}$) of each cell when the number of cells in the network change from $S$=2 to 3 and from $S$=3 to 4. This large variation in the sending probability causes a large variation in the average reception rates, as seen in Fig. 6-20b. The analysis based on the method for each cell consisted of calculating its sending probability of the CA-Msg in a time frame (6-31) and revealed a large difference between the CA-Msg reception rates of the cells at both ends of the chain structure to the other cells. In the network with three cells in the chain, the difference between the sending probability of the CA-Msg between two cells at the ends of the chain structure send and the cell at the middle is largest, 0.5 compared to $0.\overline{3}$), as calculated in Sec. 6.3.2.4.

#### 6.3.3.2 Channel Reuse Factor and Fairness-Index

The variations in the channel reuse factor and the fairness-index based on the variation in the status of users are monitored for a PLC network with 3 cells in the chain structure. Each cell consists of 20 users and each user changes its status between active and inactive in accordance with the parameters presented in Sec. 6.3.1.3. Fig. 6-21 presents the fairness-index and channel reuse factor measurement for the entire network during a short simulation run of 10,000 time frames $T_{TF}$. The average fairness-index calculated from 10 sub-runs, each lasting for 100,000 measurements during a simulation period of 100,000$T_{TS}$ is 0.993 (confidence interval (CI) is [0.991, 0.995]). This value is near to the perfect fairness (≈1) and means that the number of channels per active user ($M_i$) is quite equally distributed over the 3 cells. The average channel-reuse factor is 1.474 (CI is [1.458, 1.490]), meaning that the overall network has an average of 80×1.474=117.92 channels for data transmission.

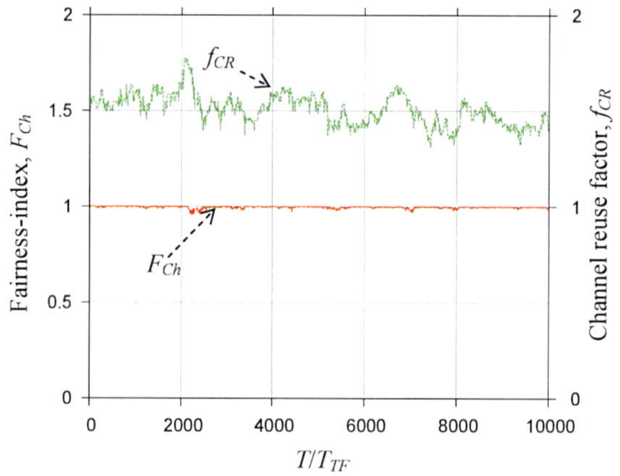


*Figure 6-21: Fairness-index and channel reuse factor measurement*

During the simulation run, no interference channel in the network was observed ($B_{Ei}$=$B_{Ci}$). By analysis of the network structure, it can also be seen that there is no situation in which two or more neighbor cells can seize the same channel released by one common neighbor cell. Therefore, the number of effective channels is equal to the sum of the numbers of channels used by all the cells.

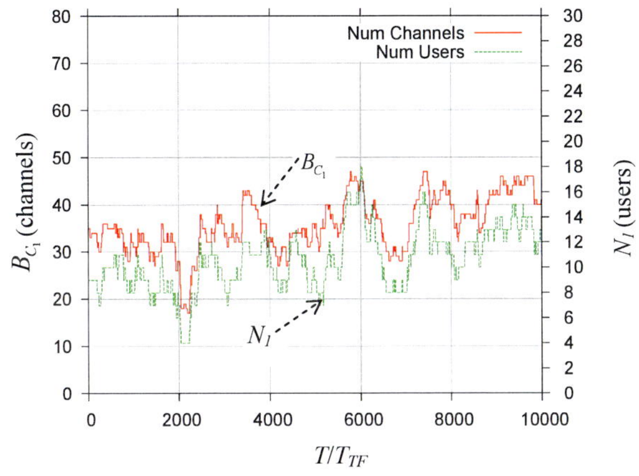


*Figure 6-22: Seized channels and active users in cell* $C_1$

For analyzing the variations of channels in a cell depending on changes in the number of active users in that cell, we present this tendency for the cell $C_1$ during the simulation time in Fig. 6-22. It is observed that the number of seized channels of the cell $C_1$ has the same trend following a change in the number of active users, which is randomly varied.

The average number of channels per user ($E\{M_i\}$) of all cells in the network is an indication for the resources reserved per active user in the whole network. During the simulation run, the variation of $E\{M_i\}$) shows an inverse trend to the variation of the average number of users per cell ($E\{N_i\}$). These variations in a period in a subrun are presented in Fig. 6-23.

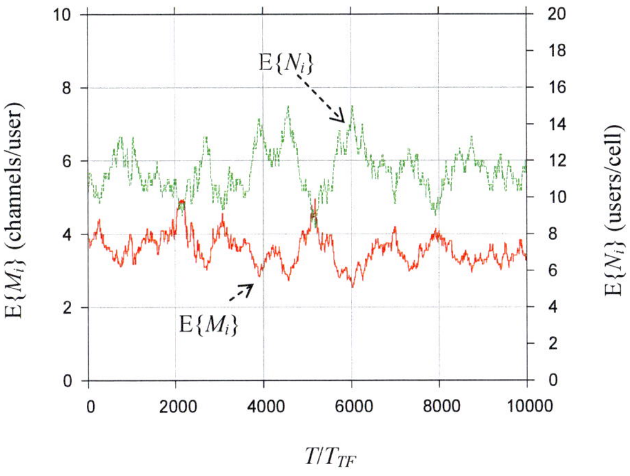


*Figure 6-23: Average channels/user and average users/cell*

From the simulation results, the average number of channels per user in the investigated network is 3.674 channels (the confidence interval is [3.603, 3.745]). Further on we will write result and this confidence interval as e.g. 3.674 [3.603, 3.745]. It indicates that on average if a user can transmit in a channel with 1Mbps, it is allocated to transmit with the average speed of 3.674 Mbps.

#### 6.3.3.3 Performance Evaluation

The performance parameters of the protocols, the sum of the number of allocated channels, the average number of channels per users, fairness-index, and the channel reuse factor are evaluated for different sizes of PLC networks. Each value of a parameter is achieved by simulating a network with the given number of cells in the chain structure and the simulation parameters presented in Sec. 6.3.1.3. The simulation results of the performance parameters are presented in Fig. 6-24. Characteristics of the parameters are analyzed with different number of PLC cells in the network.

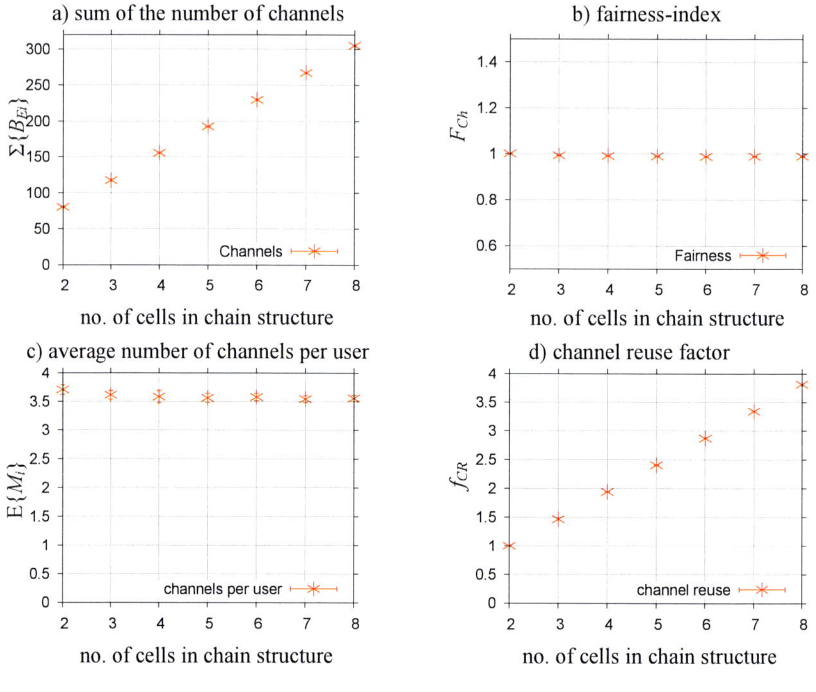


*Figure 6-24: Evaluation for the PLC networks with cells in the chain structure*

The sum of the number of allocated channels (Fig. 6-24a) is almost linearly increased with an increasing number of cells in the networks. There are no interfering channels during all simulations. The channel reuse factor also has the same characteristics of the sum of the number of allocated channels (Fig. 6-24d). For all network sizes, the protocol maintains a good fairness between users in different cells (Fig. 6-24b) in terms of the number of reserved channels per active user. The average number of reserved channels per active users is reduced slightly when the number of the cells is increased.

### 6.3.4 Scenario 2: PLC Cells in a Ring Structure

The throughput and delay for exchanging the channel information between the cells by means of the CA-Msg is evaluated, using both simulation and analysis methods. The PLC network model consists of 2 to 8 cells in the ring structure. The network with two cells is essentially not a ring structure, but it is included for reference and comparison. For the other network which has more than two cells, each cell has two neighbors. Fig. 6-25a presents the average delivery times and Fig. 6-25a shows the average throughputs of the CA-Msg between the cells in networks with different numbers of cells. The simulation parameters are described in Sec. 6.3.1.3. The analysis results are also within the confidence intervals of the equivalent simulation results.

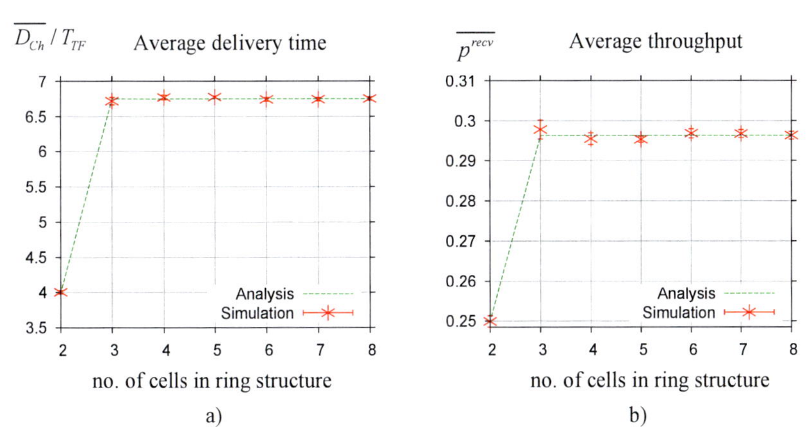


*Figure 6-25: a) Average delivery time; and b) average throughput of the CA-Msg in networks in the ring structure*

With this ring structure of the PLC network with 3 or more cells ($S \geq 3$), the average delivery time ($\overline{D_{Ch}}$, defined and calculated in (6-15) and (6-29)) and average throughput of the CA-Msg ($\overline{p^{recv}}$, calculated in (6-27)) are constants. The reason is that each cell has two neighbors, so each cell has the same sending probability of the CA-Msg in a time frame. In accordance with (6-31):

$$p_j^{send} = \frac{1}{2}\cdot\left(\frac{1}{3}+\frac{1}{3}\right) = 0.\overline{3} \tag{6-37}$$

Substitute this value to (6-27) and (6-28) it is achieved

$$\overline{D_{Ch}} = 6.75 \quad / \forall S \geq 3 \text{ and} \tag{6-38}$$

$$\overline{p^{recv}} = 8/27 = 0.\overline{296} \quad / \forall S \geq 3 \tag{6-39}$$

### 6.3.4.1 Protocol Analysis

A PLC network consisting of 8 cells, located in a ring structure in which each cell interferes with two adjacent neighbors (Fig. 6-26) is selected for the protocol performance evaluation. Simulation parameters are followed Sec. 6.3.1.3. Each cell has 20 users, and a user changes it status between inactive or active after a certain duration of time in accordance with a geometric distribution with a mean of 800 $T_{TS}$.

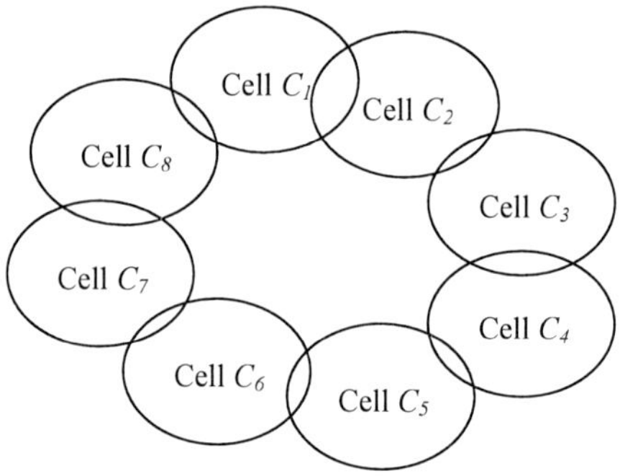


*Figure 6-26: PLC network with 8 cells in a ring structure*

Fig. 6-27 presents the fairness-index and channel-reuse factor for the entire network during the first 10.000 $T_{TF}$ of the simulation time. The average fairness-index calculated from the simulation period of 10 sub-runs, each 100.000 $T_{TF}$ is 0.988 [0.987, 0.989]. This value indicates that the number of channels per active user ($M_i$) is quite equally distributed over the PLC cells. The average channel reuse factor is 3.730 [3.716, 3.745]. This means that, the average of the actual allocated number of channels in the network during the simulation time is 80×3.730=298.4 channels.

In this network, also no interference channel has been observed during the simulation. The reason is the same as in the previous case in which there was no situation where more than two neighboring cells seized the same released channel from a common neighbor cell.

Therefore, the number of effective channels is equal to the sum of the numbers of channels used by all the cells.

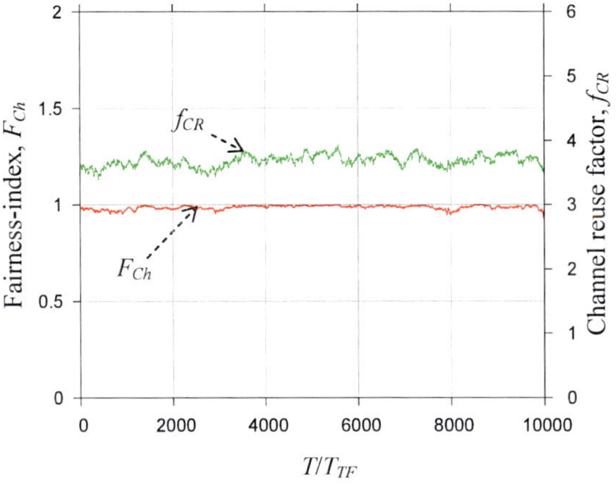


*Figure 6-27: Fairness-index and channel reuse factor measurement*

Fig. 6-28 presents the dependency of the number of allocated channels in a cell (cell $C_1$) with the number of active users in that cell. It can be seen that the number of channels that cell $C_1$ seized possesses the same trend with the change in the number of active users during the run-time. With an average number of active users of 10.890 [10.726, 11.055], cell $C_1$ is allocated an average of 37.338 [37.127, 37.548]) channels.

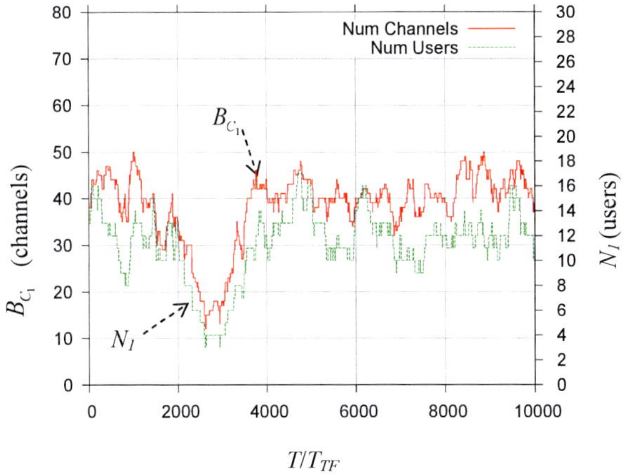


*Figure 6-28: Seized channels and active users in cell* $C_1$

The average number of channels per active user ($E\{M_i\}$) and the average number of active users per cell ($E\{N_i\}$) for the whole network are also monitored and their variations during the first 10.000$T_{TF}$ are presented in Fig. 6-29. The average number of channels per active user is inversely proportional to the average number of active users in the network. Based on the simulation results, the average number of channels per active user in this network is 3.465 [3.423, 3.506].

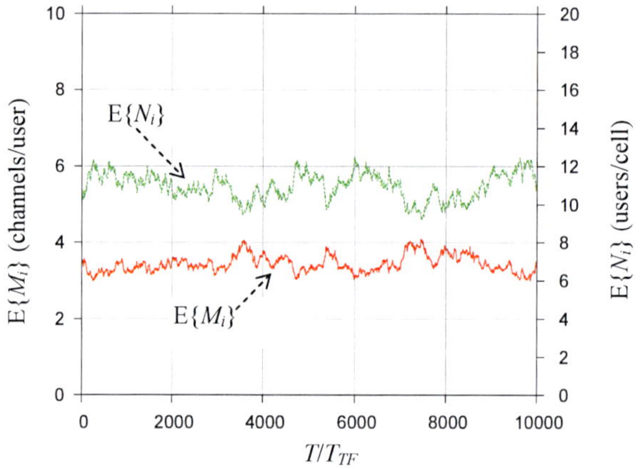


*Figure 6-29: Average channels/user and average users/cell measurement*

#### 6.3.4.2 Performance Evaluation

Diverse numbers of PLC cells in the ring structures are taken into account for the performance evaluation. Each PLC network is simulated for a long duration with the simulation parameters given in Sec. 6.3.1.3. The performance metrics including the number of allocated channels, average number of channels per users, the fairness-index, and channel reuse factor are presented in Fig. 6-30.

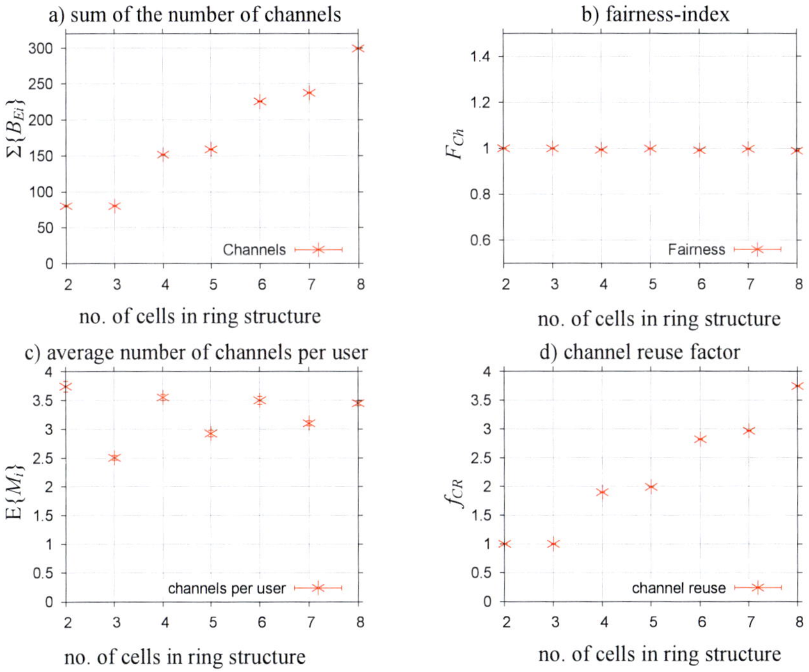


*Figure 6-30: Evaluation for the PLC networks with cells in the ring structure*

Fig. 6-30a presents the total number of allocated channels ($\Sigma\{B_{Ei}\}$) in the entire network. It can be seen that $\Sigma\{B_{Ei}\}$ increases when the number of cells in PLC network increases. However, the number of channels $\Sigma\{B_{Ei}\}$ remains approximately the same when the number of cells is an even number (in the format of $S=2x$ / $x \in \mathbb{N}$, $x>0$) and the next an uneven number (in the format of $S=2x+1$/ $x \in \mathbb{N}$, $x>0$). This feature is the same for the channel reuse factor presented in Fig. 6-30d.

From the results presented in Fig. 6-30b, it can be said that the fairness index values are almost equal to 1 for all experiment PLC network sizes. The average number of channels per active user (presented in Fig. 6-30c), however, strongly depends on the number of cells in the network, whether it is an even or an uneven number.

### 6.3.5 Scenario 3: PLC Cells in a Meshed Structure

#### 6.3.5.1 Protocol Analysis

The performance of the protocol is investigated for the PLC networks consisted of the cells in the meshed structures as the reference model presented in Sec. 4.2.1.3. Generally, a network with a meshed structure can be considered to contain chain, ring, and star parts. Because the chain and ring structures have previously been analyzed, for different PLC network sizes in this scenario, the performance metrics are evaluated for the networks in star structures.

One specific PLC network structure is selected for the protocol analysis as presented in Fig. 6-31. This network consists of 6 PLC cells, in which cells $C_1$, $C_3$, $C_6$ have one neighbor, cells $C_4$ and $C_5$ have 2 neighbors, and cell $C_2$ has 3 neighbors. The fairness index and the channel reuse factor during the first 10.000 $T_{TF}$ simulation time to perform the channel allocation protocol are presented in Fig. 6-32. The simulation results with the simulation parameters in Sec. 6.3.1.3 show that the network achieves a channel reuse factor of 2.892 [2.874, 2.910], while the fairness index is high, 0.985 [0.983, 0.988]. The channel reuse factor is almost equal to the optimal value of the channel reuse factor in which each cell is allocated half of the number of channels (optimal value is 3). The reason that the fairness index is not near 1 is that, for this network, if a channel is allocated to cell $C_1$, it can be reused 4 times (by cells $C_3$, $C_4$, $C_6$), while another channel allocated to cell $C_2$ can only be reused only 2 times (by cell $C_5$). This difference between the possibilities of channel reuse when a channel is allocated to a specific cell leads to a large difference in the numbers of channels per active user in different cells, where the channels can be reused.

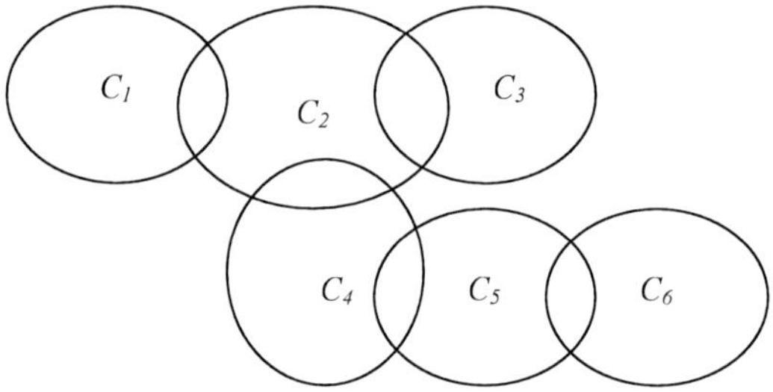


*Figure 6-31: Network with 6 PLC cells in the meshed structure*

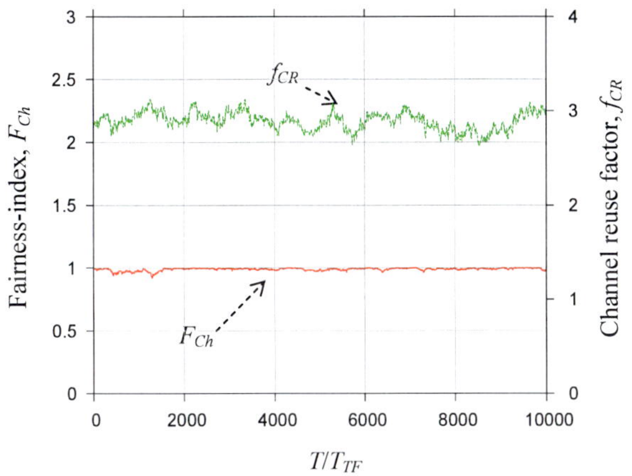


*Figure 6-32: Fairness-index and channel reuse factor during run-time*

The average number of channels per active user and the average number of users in a cell are calculated during the simulation. Their variations during the first $10.000T_{TF}$ are presented in Fig. 6-33. Similar to the results for the other networks previously presented, the average number of channels per active user is inversely proportional to the average number of active users.

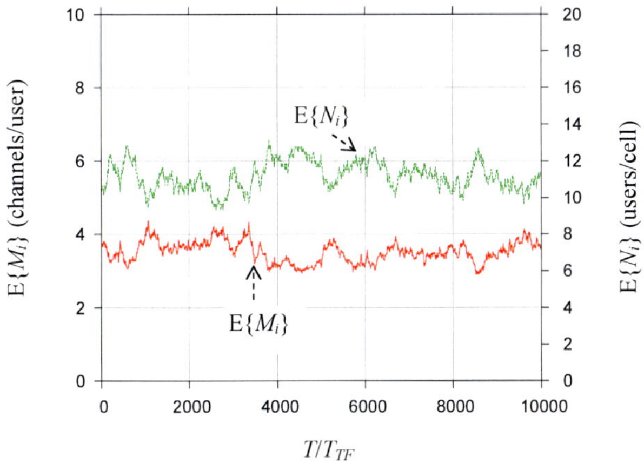


*Figure 6-33: Average channels/user and average users/cell of the network with 6 PLC cells in the meshed structure*

### 6.3.5.2 Performance Evaluation

For the analysis of the channel allocation protocol with different PLC network sizes considering the meshed structure, a star network structure has been selected. As already mentioned, a meshed network may comprise of several parts; each can be a chain, ring, or star structures. Therefore, the performance of the proposed DICRA protocol in PLC networks with star structures are evaluated. In this network structure, each cell is connected to only one cell in the middle of the network. This is the typical structure of a PLC network in which an access cell is connected to several separate inhome cells. This network structure consists of one cell ($C_1$) in the middle of the network and the other cells ($C_2..C_S$) can exchange their CA-Msg only with the middle cell ($C_1$), as presented in Fig. 6-34.

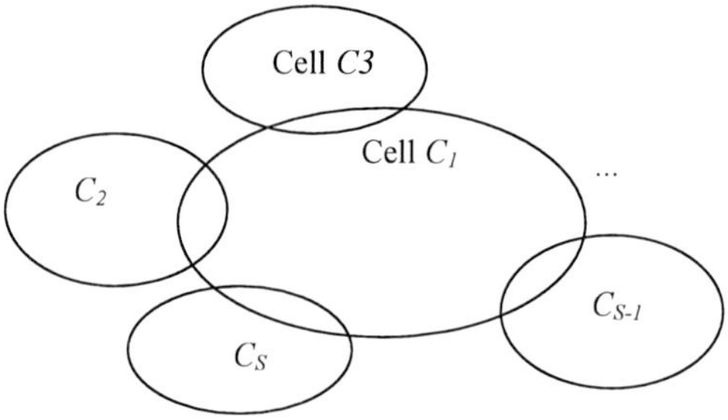


*Figure 6-34: PLC network in a star structure with S cells*

The average delivery time and average throughput of the CA-Msg between the cells are presented in Fig. 6-35. In this figure, both simulation results with the simulation parameters in Sec. 6.3.1.3 and calculation results are presented. It can be seen that the average delivery time and the average throughput increase when the number of PLC cells in the network increases. In accordance with A.2.2:

$$\lim_{S\to\infty}(\overline{D_{Ch}}) = \infty \text{ and} \tag{6-40}$$

$$\lim_{S\to\infty}(\overline{p^{recv}}) = 0.5 \tag{6-41}$$

This demonstrates that, when there are several PLC cells in the star network, the CA-Msg can be exchanged with the average throughput at each cell of 0.5. However the delay in the transmission of the CA-Msg messages from the cells around to the middle cell ($C_1$) is significantly long.

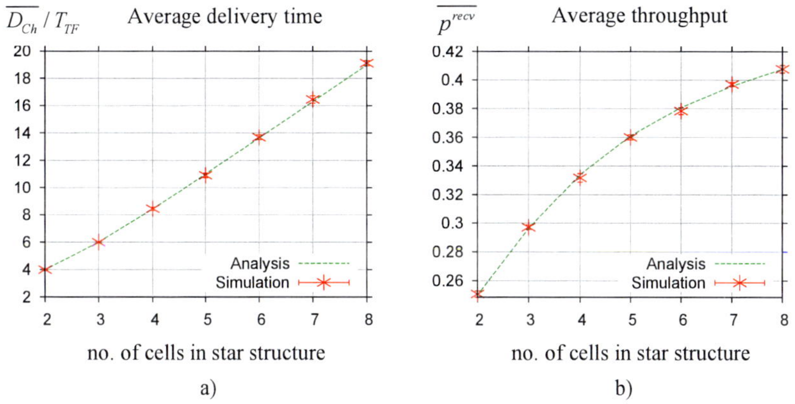


*Figure 6-35: a) Average delivery time; and b) average throughput of the CA-Msg in the networks in the star structure*

The other performance metrics of the protocol are simulated for the networks with different sizes and presented in Fig. 6-36. Based on the simulation results presented in Fig. 6-36a, the sum of the number of allocated channels in the networks increases when the number of cells increases. For this network, a channel allocated to $C_1$ is not reused but a channel allocated to $C_2$ can be reused by all the other cells except the cell $C_1$. This is similar to the channel reuse factor (Fig. 6-36d) in which the value is proportional to the number of cells in star networks.

The protocol maintains high fairness indices (close to 1 as presented in Fig. 6-36b) meaning that the numbers of allocated channels per active user in different cells is similar. It can also be seen that the average number of channels per active user essentially does not depend on the size of the PLC network (Fig. 6-36c).

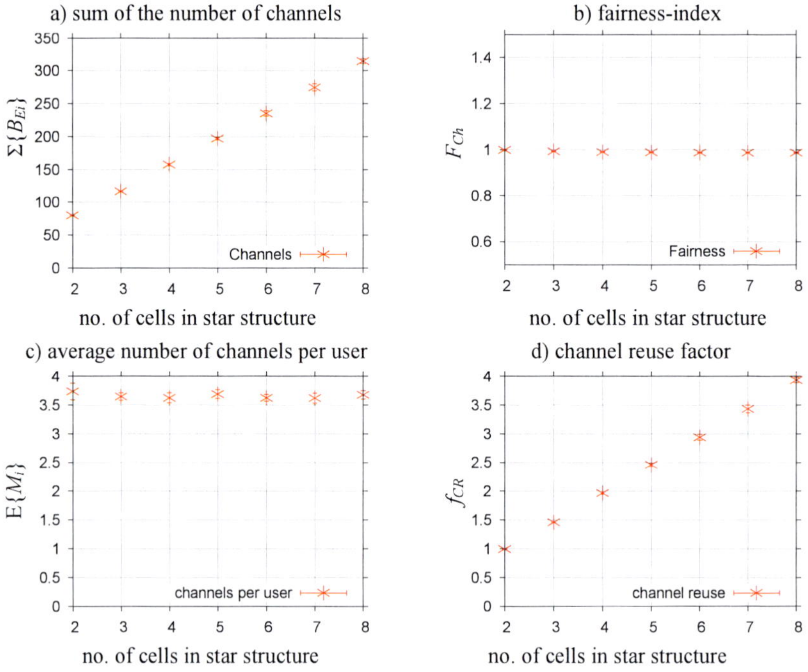


*Figure 6-36: Evaluation for the PLC networks with cells in the star structure*

### 6.3.6 Comparison with IPP

Another experiment is carried out to compare the performance of the proposed DICRA protocol with the IPP (Inter PHY Protocol [HomePa07] [GallKu09], for the network presented in Fig. 6-31. Applying the IPP protocol, each BS broadcasts its presence only to its neighbors. When a cell detects available neighbors, it has to adjust its transmission resources, in term of time slots, accordingly. During run-time, if a BS detects one neighboring cell, this cell has to reduce its transmission resources to half. If it detects two neighbor cells, this cell has to reduce its transmission resources to a third, and so on for additional neighbors.

Two performance parameters, the average number of channels per active user ($M$ value) and fairness-index between the users, are selected for the comparison. The two protocols are simulated parallel with the same variation of the users in different PLC cells. For the IPP protocol, each allocation time slot is called channel and there is no channel reallocation because the network structure is constant during the simulation time. The variations in the simulated results are presented in Fig. 6-37. $F_{Ch}$ and $E\{M_i\}$ denote the fairness index and average $M$ value of the proposed DICRA protocol and $F_{Ch}^{IPP}$ and $E\{M^{IPP}\}$ are the corresponding results achieved from the IPP protocol.

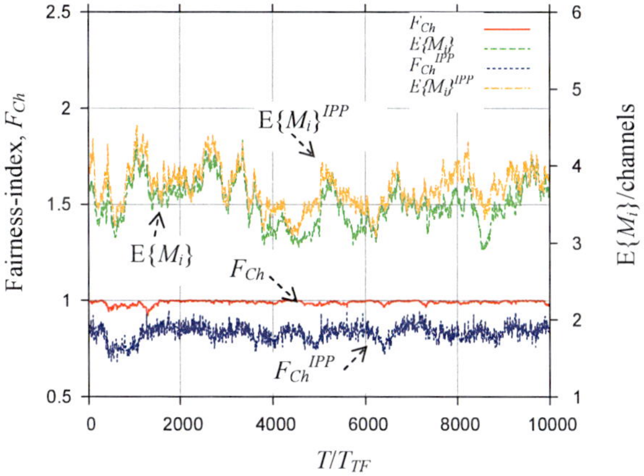


*Figure 6-37: A comparison with the IPP protocol*

In accordance with the simulation results, the average number of channels per active user achieved from both protocols has the same tendency during the runtime. The average values calculated during a long simulation with the simulation parameters in Sec. 6.3.1.3 are 3.569 [3.507, 3.632] and 3.802 [3.748, 3.855] for the proposed DICRA protocol and the IPP, respectively. This means that the average number of channels per active user using IPP is about 6.5% higher than the number of channels per active user using the proposed DICRA protocol. The channel reuse factor with the proposed protocol is 2.892 [2.874, 2.910], while the IPP is constant at 3.

However, the advantage of the proposed protocol compared to the IPP is that the proposed protocol allows the channels to be reallocated between PLC cells. The dynamic reallocation of the channels is necessary for the multi-cell PLC system, because it eliminates the case in which users in some PLC cells do not have enough resources for their transmissions while channels in the other PLC cells are not in use. The reallocation reduces the difference between the numbers of channels per active user in different cells. Therefore, the fairness-index achieved by the proposed protocol fairness is high, 0.985 [0.983, 0.988], and it is much higher than that (average 0.838 [0.836, 0.844]) achieved by the IPP protocol. The standard deviations when calculating the statistical results of the number of channels per active user, which measures the spread of the $M$ values in different cells from the value of $E\{M_i\}$ and $E\{M_i\}^{IPP}$, are compared. The population standard deviation of the $M$ value in different cells using IPP is more than three times the $M$ value achieved by the proposed DICRA protocol.

## 6.4 Summary

The proposed channel allocation protocol (DICRA) allows for dynamic re-allocation of the channels between the PLC cells in the network during run-time. The objectives of the re-allocation of the channels are to maintain fairness between the users in different cells in terms of the number of channels per active user, while maximizing the channel reuse factor, and to avoid the usage of the same channels in the neighboring cells.

This protocol requires modifications in the base station of each cell. These modifications would allow a base station to exchange channel information with the other cells via the so-called channel announcement message (CA-Msg). One time slot with the duration of a OFDM symbol is reserved for sending this message between the base stations at the beginning of every time frame (*TF*). The transmission of the CA-Msg is based on the in-line interference characteristic of the PLC network: when two cells are interfering each other, they share at least the same electrical line.

The protocol is evaluated by means of simulation and analytic methods. In most of the referenced PLC networks, the protocol achieves near-perfect fairness between the users in different cells, the channel reuse factor is high and there are no occurrences of interfering channels.

Generally, the DICRA protocol can be applied to both types of the PLC access cells and PLC inhome cells. Since it is designed to work in the distributed fashion, a cell needs only the channel information from its neighbor to locally determine its reserved channels. Based on the analysis of the results, the performance of the protocol strongly depends on the number of neighbors of a cell, but not on the size of the PLC network.

# Chapter 7 Optimized Intra-Cell-PLC Scheduling

In this chapter, we propose and investigate scheduling algorithms for a single cell in an entire PLC system. The algorithms allow the base station to allocate transmission resources efficiently to support QoS guarantees for different services to end users. The functionalities of the MAC- and PHY-layer in the OPERA Specification [Opera06] are used as the base for the investigation. The OPERA Specification is reviewed in Sec. 2.3.

In each PLC cell, the base station (BS) collects information of the transmission rates in the physical layer via the so-called channel measurement procedure. This procedure is regularly triggered to update the knowledge about the channel qualities of all the connections to all the users in the network. After the measurement, the data rates of the connections are reported to the BS to generate the overall knowledge of the logical connections. Disturbances are reflected via the measured data rate. If disturbances in a logical connection are high, a more robust modulation scheme or higher redundancy for the error correction may be used and thus may reduce the data rate. The term PLC network is used in this chapter, also implying a network with a single PLC cell.

It is assumed that the network topologies, channel qualities, available resources, and resource requests from all users are known by the BS. The task of the scheduler is to assign the available resources to each user in accordance with the requests in the entire PLC cell. The problem occurs when there are not enough resources to fulfill all requests from the users. In order to do so, the resource allocation problem is formulated into an optimization problem. The problem is solved by deterministic algorithms and a heuristic algorithm, and their performances are compared to the round robin scheduler. A so-called adaptation strategy is proposed to speed up the solution time when applying the heuristic algorithm. The problem is also investigated with the utilization of the OFDMA access scheme.

Detailed channel models have been studied for channel characteristics of the PLC connections, e.g. as presented in [Dost01] and [ToneZh09]. This chapter, however, focuses on the scheduling technique in which the information from the channel is provided by the physical layer through the current transmission data rate of each user. Therefore, a simplified model is used to represent the physical layer.

## 7.1 Scheduling with TDMA

The scheduler is a function implemented in the base station (BS) in order to manage the available transmission resources and grant access to the medium by its slaves in both uplink and downlink transmissions. A good sharing strategy has to ensure the effectiveness of the granted service and fairness between multiple users. One problem formulation scheme is presented in [DoLe08], in which a complex economical model is used for calculating the utilization of the granted services for a user.

### 7.1.1 Logical Connection and Traffic Flow

When a user connects to the network, a connection from that user to the BS of the PLC-cell is established. In order to provide the QoS guarantee for a communication service, a certain amount of the transmission resources will be assigned to each transmission flow. The relationship of the physical medium, logical connections, and traffic flows between a user and its master (BS) is presented in Fig. 7-1. Over a logical connection between a user and its base station, several traffic flows can be granted. Each service is classified into a specific service class in which a group of QoS parameters, e.g. throughput, latency, transmission reliability, etc., are defined. For example, the throughput requirement for VoIP, according to [ITU-G.711], is 64kbps with an end-to-end delay less than 150ms; a video conference flow requires a throughput of 128kbps.

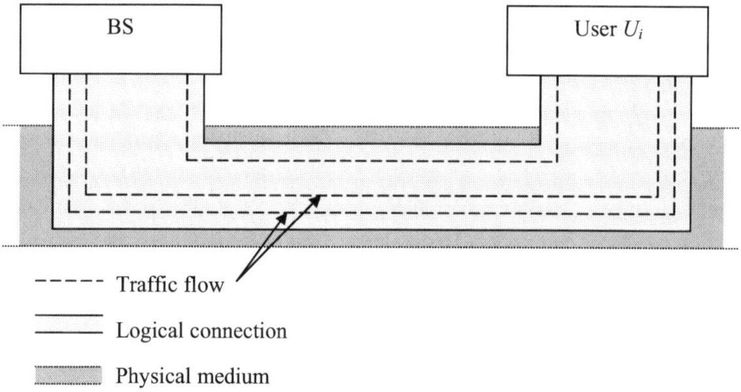


*Figure 7-1: Relationship between flow, logical connection, and physical medium*

The resource manager uses the centralized scheduling principle (as presented in Sec. 3.3.1.3): the BS organizes the resources for all traffic flows. In accordance to this scheme, if a service is initialized by the BS, the request will be passed to the scheduler. If a new service is initialized from a user, the user sends a request to its BS and waits for a new bandwidth assignment. After the scheduler allocates additional bandwidth for the new service, the BS informs the user about the new bandwidth allocation. After that, the transmission of the data from the new service may start.

Both time division repeaters (TDR) and frequency division repeaters (FDR) are used to extend the coverage of a PLC cell. However, extending the coverage between a master and a user by using repeaters either drops the performance of the logical connection comparing to each single connection or requires more transmission resources. In order to guarantee the throughput for a traffic flow, more transmission resources are necessary if repeaters are used. For the investigation, we assume that the PLC system uses TDR repeaters. With the TDMA mechanism in accordance with the specification in [Opera06], the maximum transmission delay is limited by the duration of the time frame. Guaranteeing the throughput of the traffic flows is one objective of the scheduling. In a PLC link with a specific modulation scheme, if the transmission duration of a service in every time frame is longer the throughput of that service is higher.

### 7.1.2 Transmission Resources

In order to provide a specific throughput for a service, an amount of transmission resources has to be reserved. In the TDMA system, the transmission resources are defined as the duration that a device can access the network in every time frame. For each service of a user, the transmission resources are calculated based on the actual data rate over the logical connection between the user and its master.

#### 7.1.2.1 Effective Data Rate

The effective data rate is defined as the ratio of the amount of transmitted data and the necessary transmission duration over a logical link between a user and its master. When there is no repeater in between, the effective data rate is the physical transmission data rate of the connection. Assuming that there are $r$ repeaters between a user $U_i$ and its master (BS), Fig. 7-2 demonstrates the logical connection between a user and its master before and after the conversion. The method of conversion is presented in [DoLe08]. The resources using for the guard intervals depend on the number of repeaters. These resources are eliminated from the calculation because it is assumed that they are directly reserved for the frame builder in the MAC layer.

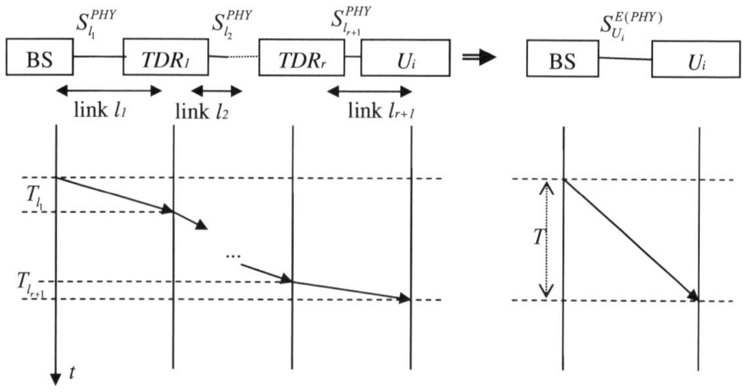


*Figure 7-2: Effective data rate conversion*

According to [Opera06], parallel transmissions inside a PLC cell are not allowed. This means that the transmission of a packet from BS to the $U_i$ is sequentially repeated from $TDR_1$ to $TDR_r$. Considering only the transmission times over the links, the duration for the transmission of a packet of length $L$ from BS to user $U_i$ is

$$T = \sum_{i=1}^{r+1} T_{l_i} = \sum_{i=1}^{r+1} \frac{L}{S_{l_i}^{PHY}}, \tag{7-1}$$

in which $S_{l_i}^{PHY}$ and $T_{l_i}$ are the data rate and the transmission time over a link $l_i$. With the transmission duration calculated in (7-1), the effective data rate of the logical link between BS and $U_i$ is

$$S_{U_i}^{E(PHY)} = \frac{L}{T} = \frac{1}{\sum_{i=1}^{r+1} \frac{1}{S_{l_i}^{PHY}}} \tag{7-2}$$

If the MAC layer allows parallel transmissions in the link $l_i$ and link $l_j$ when these transmissions do not disturb each other, more data may be transmitted between the BS and $U_i$; this is a kind of pipelined transport. In this case, the effective data rate is higher than the one in (7-2). Equation (7-2) can be considered as a calculation for the worst case of a multiple link connection in which parallel transmissions are not allowed.

### 7.1.2.2 Resources Estimation

Using the effective data rate of the connections to its users, the scheduler in the BS divides the available resources to fulfill the requirements for different services from different users. With a granted time duration $T_{i,j}$ to a service $q_j$ in a user $U_i$ in every time frame $T_{TF}$, the throughput of the service to that user is

$$\beta_{i,j} = S_{U_i}^{E(PHY)} \cdot \frac{T_{i,j}}{T_{TF}} \tag{7-3}$$

We can rephrase (7-3) to calculate the transmission resources, defined as the transmission duration in every time frame, needed for the service $q_j$ as

$$T_{i,j} = T_{TF} \cdot \frac{\beta_{i,j}}{S_{U_i}^{E(PHY)}} \tag{7-4}$$

An example is presented in Fig. 7-3 for a network which consists of a BS and three users. The transmissions in a time frame are organized for the users sequentially. Considering the priorities of the services from different users, the portions each time frame are firstly allocated to users $U_1$ and $U_2$ who have VoIP and Video services, each requiring a fixed amount of throughput. The rest of the transmission time in the time frame is then allocated to the FTP and HTTP services in $U_1$ and $U_3$ which are classified as best effort services. The fragmentations in a time frame presented in Fig. 7-3 demonstrate the allocated transmission durations of the users, grouped for both uplink and downlink transmissions. It is also demonstrated that for the transmissions of each user, an amount of resources for signaling is necessary.

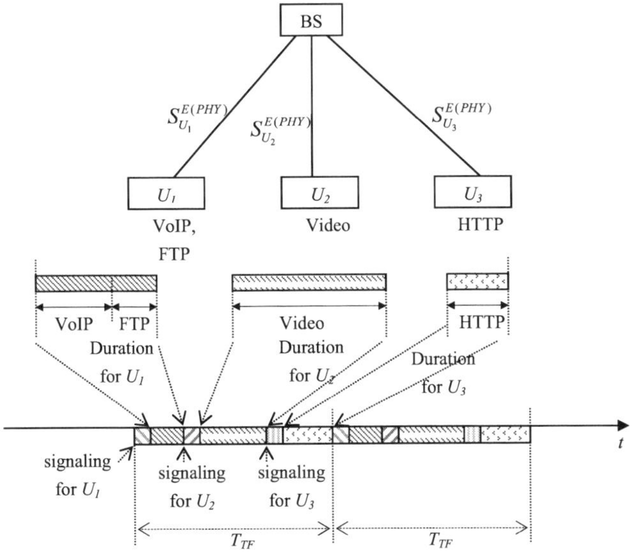


*Figure 7-3: Resource allocation in time frames*

This chapter concentrates on scheduling for services that require a QoS guarantee. After the resources have been assigned to the services with QoS guarantee, the rest of the available resources are uniformly assigned to all best effort (BE) services. These BS services do not require QoS guarantee.

### 7.1.3 Scheduling as an Optimization Problem

In order to formulate the scheduling of a PLC cell into an optimization problem, the following notations are used:

$N$: number of users in the investigated PLC cell

$U_i$: user with index $i$ ($i$=1..$N$)

$Q$: number of provided service classes

$q_j$: service class $j$ ($j$=1..$Q$)

$b_j$: benefit of service $j$

$R_{i,j}$ : transmission resources for service $q_j$ from user $U_i$

$T_{i,j}$ : transmission duration for service $q_j$ from user $U_i$

$B_{Ui}$: benefit of user $U_i$

$R_{Ui}$: transmission resources for the transmission of user $U_i$

$R_{U_i}^{OH}$ : overhead for the transmission of user $U_i$

$B_U$: benefit of a scheduling solution

$F_U$: fairness of a scheduling solution

$G$ : fitness of a scheduling solution

$\Delta G_{i,j}^{+}$ : fitness gained by assigning resources to service $q_j$ of user $U_i$,

$y_{i,j}$: number of requested services of type $q_j$ from user $U_i$

**Y**: request matrix, $\mathbf{Y}=\{y_{i,j}\}$

$x_{i,j}$: number of granted services $q_j$ to user $U_i$

**X**: grant matrix, $\mathbf{X}=\{x_{i,j}\}$

$\beta_{i,j}$ : throughput of the service $q_j$ in user $U_i$

$T_{i,j,f}$: transmission time for service $q_j$ of user $U_i$ in channel $f$

**T**: transmission time allocation matrix

$B_C$: number of channels in the PLC cell

For the scheduling problem, we consider a PLC cell with a base station and $N$ users, from $U_1$ to $U_N$. The network provides services in $Q$ service classes, and they are decreasingly sorted by their benefits, from $q_1$ to $q_Q$. The benefit of a transmission resource sharing solution represents how well the scheduler uses the available resources, called intra-cell benefit. It is calculated based on the benefit that each user achieves when its service request is granted. For example, a user may give a higher benefit to a granted VoIP connection than for a data connection with the same transmission throughput.

As mentioned before, the PLC network provides a QoS guarantee for four service classes. The service parameters, throughput, delay, as well as the benefit $b_j$ achieved when a service $q_j$ is granted are presented in Tab. 7-1.

| Service Class | Throughput $\beta_j$ | Delay | Benefit $b_j$ | Example |
|---|---|---|---|---|
| $q_1$ | CBR: 64kbps | 30ms | 8 | VoIP |
| $q_2$ | VBR: 128kbps | 60ms | 4 | Video, Game |
| $q_3$ | VBR: 32kbps | 120ms | 2 | E-commerce |
| $q_4$ | ABR: 8kbps | 240ms | 1 | Web traffic |

*Table 7-1: Services' benefit and QoS requirements*

At a point in time, a user $U_i$ may request one or several kinds of services, and even several instances of a service. The number of services of type $q_j$ that a user $U_i$ requests is denoted by $y_{i,j}$. Due to the limitation of available transmission resources, the scheduler may grant this user a different number of that services ($x_{i,j}$). These requests are collected and stored in the base station in the form of a request matrix. Fig. 7-4 presents an example of the service requests and service endings during run-time. The request matrix is updated at the end of each time frame by collecting all this information during run-time. After the scheduler decides, the corresponding users are informed of the granted services. Therefore, there are two matrices stored in the base station:

- request matrix $\mathbf{Y}=\{y_{i,j}\}$ and
- grant matrix $\mathbf{X}=\{x_{i,j}\}$

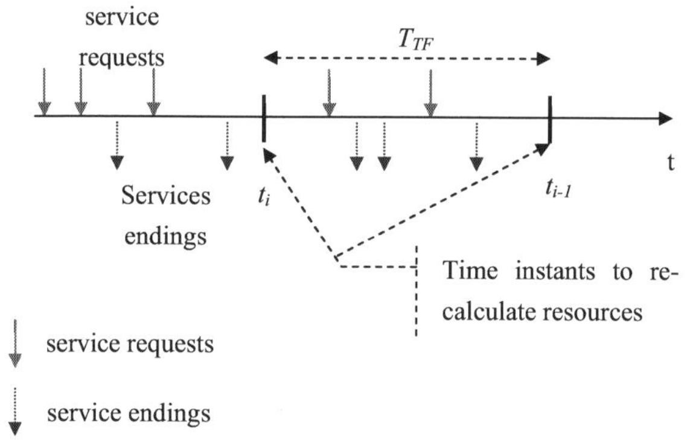


*Figure 7-4: Updates the request matrix* **Y** *in a base station*

Each user can request multiple instances of a service, therefore: $0 \leq y_{i,j} / y_{i,j} \in \{\mathbb{N}\}$. Also, the scheduler does not grant a service to a user without a service request, thus $0 \leq x_{ij} \leq y_{i,j} / x_{i,j} \in \{\mathbb{N}\}$. In summary, the elements of these matrices should follow the conditions:

$$0 \leq y_{i,j},\ y_{i,j} \in \{\mathbb{N}\} \text{ and } 0 \leq x_{ij} \leq y_{i,j},\ x_{i,j} \in \{\mathbb{N}\} \tag{7-5}$$

#### 7.1.3.1 Intra-Cell Benefit

Denoted the benefit of all granted services for a user $U_i$ by $B_{Ui}$, the value of the user's benefit is calculated by summing up all the benefits from individual granted services.

$$B_{Ui} = \sum_{j=1}^{Q} \left( x_{i,j} \cdot b_j \right) \tag{7-6}$$

The benefit of the entire network is the sum of all the benefits from all users. With $N$ users in the PLC cell, the intra-cell benefit is defined as

$$B_U = \sum_{i=1}^{N} B_{Ui} = \sum_{i=1}^{N} \sum_{j=1}^{Q} \left( x_{i,j} \cdot q_j \right) \tag{7-7}$$

The network benefit also represents the effectiveness of the scheduling to grant the available resources to the requested services.

#### 7.1.3.2 Intra-Cell Fairness

Several fairness definitions between users have been studies in different kind of networks, e.g. as presented in [GambYu04], [RaduBo07], [KellMa98], or [VandFa00]. For our problem formulation, we define the fairness derived from the definition of the proportional fairness: the scheduling solution is fair if the users have the same ratio between the number of granted services and the number of requested services. The calculation to evaluate if these ratios are the same for all users is inspired by the Jain's fairness index calculation [JainCh84].

Applying to the PLC cell, for all $\sum_{j=1}^{Q} y_{i,j} > 0$:

$$F_U = \frac{\left(\sum_{i=1}^{N}\left(\sum_{j=1}^{Q} x_{i,j} \Big/ \sum_{j=1}^{Q} y_{i,j}\right)\right)^2}{N \cdot \sum_{i=1}^{n}\left(\sum_{j=1}^{Q} x_{i,j} \Big/ \sum_{j=1}^{Q} y_{i,j}\right)^2} \tag{7-8}$$

From (7-8)

$$F_U = \begin{cases} 1 & \text{if } \sum_{j=1}^{Q} x_{i,j} \Big/ \sum_{j=1}^{Q} y_{i,j} \text{ is the same for all users} \\ <1 & \text{otherwise} \end{cases} \tag{7-9}$$

It is said that the resource allocation solution is completely fair if the sum of the ratio of the granted services per requested services is the same for all users.

#### 7.1.3.3 Resource Utilization

In order to satisfy the QoS requirements for a service $q_j$ of user $U_i$, we assume that an amount of transmission resources $R_{i,j}$ is necessary. With the assumption of utilizing the TDMA scheme, the transmission resources are defined as portions of time in a time frame ($T_{i,j}$) for the transmission of that service. Therefore, for our resource definition in the pure TDMA system: $R_{i,j}=T_{i,j}$. Individual users may require a different amount of resources for satisfying the same kind of service because transmission data rates between the users and the base station (BS) are usually not the same. From (7-4), the transmission duration necessary to be allocated for service $q_j$ by the user $U_i$ is calculated by

$$R_{i,j} = T_{i,j} = T_{TF} \frac{\beta_{i,j}}{S_{U_i}^{E(PHY)}} \tag{7-10}$$

The total resources allocated to a user $U_i$ is summed up from all granted services

$$R_{Ui} = \sum_{j=1}^{Q} \left(R_{i,j} \cdot x_{i,j}\right) \tag{7-11}$$

For the entire PLC network, the total allocated resources $R_U$ and the necessary overhead for signaling all transmissions to all users ($R^{OH}$) must be smaller or equal to the total available resources in the PLC cell ($R_{Max}$). Provided the overhead necessary for organizing the transmission of user $U_i$ is $R_{U_i}^{OH}$, the boundary condition of the resource allocation is

$$R_U = \sum_{i=1}^{N}(R_{U_i} + R_{U_i}^{OH}) = \sum_{i=1}^{N}\left[\sum_{j=1}^{Q}\left(R_{i,j}\cdot x_{i,j}\right) + R_{U_i}^{OH}\right] \leq R_{Max} \tag{7-12}$$

In summary, the intra PLC cell resource allocation problem is formulated into a two-objective optimization problem. This is to find a grant matrix **X** in order to

*maximize* $B_U$ in (7-7) and

*maximize* $F_U$ in (7-8)

such that

(7-5) and (7-12) are fulfilled.

For solving this optimization problem with two objectives, multi-objective solvers can be used. The other method is to convert this into a single objective optimization problem, e.g. by means of a weight aggregating function [SumaKu06]. By doing so, the problem can also be solved by a corresponding single objective solver. For this purpose, two parameters, $\alpha_1$ and $\alpha_2$, are used for specifying the importance of the two objectives. The single objective, called fitness, is then defined from the two objectives calculated in (7-7) and (7-8) by

$$G = \alpha_1 B_U + \alpha_2 F_U \tag{7-13}$$

Since the objectives have different ranges, i.e. $B_U \in [0; B_U^{\max}]$ and $F_U \in [0;1]$, it is necessary to scale $F_U$ with the factor of $B_U^{\max}$ to obtain an equal weight for the objectives. However, since the value of $B_U^{\max}$ is not known before the problem is solved, the benefit is used for scaling the $F_U$. Therefore, the parameters are $\alpha_1 = 1$ and $\alpha_2 = B_U$, and (7-13) can be rewritten as

$$G = B_U + B_U \cdot F_U = B_U(1 + F_U) \tag{7-14}$$

The scheduling is formulated as a single objective optimization:

*maximize* $G$ in (7-14)

such that

(7-5) and (7-12) are fulfilled.

The algorithms that have been used for solving the optimization problems in [DoLe08] and [DoLe09] are adapted to solve this scheduling problem. The performances of the algorithms are improved by means of corresponding procedures.

### 7.1.4 Deterministic Algorithms

In order to solve the resource allocation problem formulated in Sec. 7.1.3, an exhaustive search (ES) is implemented, in which all possibilities of the combination of the variables are evaluated. Theoretically, it allows one to find the global optimal solution. Due to the long calculation time of the exhaustive search, two other methods, namely round robin based resource distribution (RRS) and benefit based selection algorithm (BBS), are also investigated. With some approximations in these algorithms, a solution of the problem can be found in a relatively short calculation time.

### 7.1.4.1 Exhaustive Search (ES)

The exhaustive search finds grant matrix **X** from the request matrix **Y** by examining $N \times Q$ variables. The flowchart of the exhaustive search is presented in Fig. 7-5. The search algorithm uses the recursive implementation technique for optimizing the code and execution time. Since every variable is calculated, the time complexity is $O(\prod_{i=1}^{N}\prod_{j=1}^{Q} y_{i,j})$. Due to the high computation requirements, this method may be used only with a small size network.

```
Begin
        Init: i:=1;j:=1;x_i,j:=0; G:=0;G_Max=0;R=0;
        Call function ES(1,0);
End
Function ES(i,j):
if i<N, j<Q
        for (int x_ij_tmp=0;x_ij_tmp<=y_ij;x_ij_tmp++)
                x_i,j:=x_ij_tmp;
                Call function ES(i,j+1); //recursive with same i, next j
        end for
elseif j==Q
        for (int x_ij_tmp=0;x_ij_tmp<=y_ij;x_ij_tmp++)
                x_i,j:=x_ij_tmp;
                Call function ES(i+1,0); //reductive with next i, begin j
        end for
else //i==N, j==Q
        calc. G
        if (G_max<G) and (R≤R_Max)
                G_max=G
                update solution (X)
        endif
End of function ES;
```

*Figure 7-5: Flowchart of the exhaustive search algorithm*

#### 7.1.4.2 Round Robin Based Resource Distribution (RRS)

The resources are uniformly distributed to all users who request resources, regardless of their amount of traffic requirements. The users are served in the round-robin principle (RRS): the scheduler consecutively receives data from the queues of each user until it fulfills the reserved transmission resources or there is no more data to be transmitted. By utilizing the *sign*() function to indicate if a user requests data or not

$$sgn\left(\sum_{j=1}^{Q} y_{i,j}\right) = \begin{cases} 1 & \text{if user } U_i \text{ requests at least one service} \\ 0 & \text{otherwise} \end{cases} \tag{7-15}$$

Therefore, each user who requests at least one service is reserved a certain amount of resources:

$$R_{Ui} = \frac{R_{Max} - \sum_{i=1}^{N} R_{U_i}^{OH}}{\sum_{i=1}^{N} sgn\left(\sum_{j=1}^{Q} y_{i,j}\right)} \tag{7-16}$$

For each user, the scheduler allocates the resources to the services with the highest priority first until all resources are consumed or there are no further requests. The time complexity of this method is $O(N{\cdot}S)$. This method provides very fast decisions but it has the disadvantage that the solution is not optimized, by means of both benefit and fairness.

#### 7.1.4.3 Benefit Based Selection Algorithm (BBS)

The benefit based selection algorithm utilizes the mechanism inspired by the max-min fairness algorithm presented in [RaduBo07]. Let $\Delta G_{i,j}^{+}$ be an additional fitness gained by assigning resources to service $q_j$ to the user $U_i$, which consumes resources $R_{i,j}$. This algorithm contains the following steps:

- Step 1: Calculate $\frac{\Delta G_{i,j}^{+}}{R_{i,j}}$ for all possible assignments for all users and all services
- Step 2: Assign resources $R_{i,j}$ to the request which achieves the highest $\frac{\Delta G_{i,j}^{+}}{R_{i,j}}$
- Step 3: Repeat Step 1 and Step 2 until it meets one of the following conditions: all service requests are granted or there are no more resources.

The flowchart of the algorithm is presented in Fig. 7-6.

```
Begin
        Init: x_{i,j}:=0; R=0;
Main_Loop:
        Init: i:=1;j:=1; G:=0; can_add_ser=0;
        For all i,j
                If x_{i,j}< y_{i,j}
                        ΔG^+_{i,j} = G(x_{i,j}+1) - G(x_{i,j})
                        If (ΔG^+_{i,j}/R_{i,j} > ΔG^{+,Max}_{i,j}/R_{i,j}) and (R+R_{i,j})<R_{Max})
                                can_add_ser =1
                                i_{ud}=i; j_{ud}=j
                                ΔG^{+,Max}_{i,j}/R_{i,j} = ΔG^+_{i,j}/R_{i,j}
                        endif
                Endif
        Endfor
        If (can_add_ser=1) then //one service can be added
                x_{i_{ud} j_{ud}} = x_{i_{ud} j_{ud}} + 1
                R=R+R_{i,j}
                Goto Main_Loop
        End if
End
```

*Figure 7-6: Flowchart of the benefit based selection algorithm*

Let $d$ be the total number of requirements: $d=\sum_{i=1}^{N}\sum_{j=1}^{Q} y_{i,j}$ . In the worst case, the algorithm runs for $d+(d-1)+\ldots+2+1=\frac{d(d+1)}{2}$ times of benefit calculations for all granted services. Eliminating the time for comparison, the time complexity of this method is $O(\frac{d(d+1)}{2})$.

### 7.1.5 Heuristic Algorithm (HA)

There are several studies and proposals for solving the optimization problems utilizing heuristic algorithms. In [Torr08], several heuristic algorithms are reviewed and compared by means of the execution time and the complexity when solving an optimization problem. Regarding computational requirements and quality of solution, Simulated Annealing (SA)

[KirkGe83] is selected for solving a single objective optimization problem. Therefore, a heuristic algorithm derived from the principle of the SA algorithm is proposed. The proposed heuristic algorithm (HA) rearranges the resources between different service requests when the resources are under a threshold ($R_{Thres}$) in order to avoid the so-called local optimum [KirkGe83].

As the first step, the algorithm finds a feasible solution by means of the RRS algorithm. After that, the services are granted by utilizing the benefit based selection (BBS) algorithm. When the resources are lower than or equal to $R_{Thres}$, the resources are taken by randomly removing services from a randomly selected user. The procedure is repeated for a given number of iterations. The search also stops if the calculation time is greater than a given limitation. In summary, the HA algorithm is presented in the following steps:

- *Step 1:* Initialize the solution by means of the RRS
- *Step 2:* Upgrade the solution (add services)
  - Calculate $\frac{\Delta G_{i,j}^{+}}{R_{i,j}}$ for all the possible upgrades
  - Do upgrades with the highest value $\frac{\Delta G_{i,j}^{+,Max}}{R_{i,j}}$
- *Step 3:* Downgrade the solution (remove services)
  - Do one or some random downgrades (remove a random service from a random user from the allocated services) when the rest of the resources is smaller than or equal to a given $R_{Thres}$
- *Step 4:* Repeat Step 2 and Step 3 until the stop criterion is met. One repetition of Step 2 and Step 3 is called an iteration.

The calculation for the time complexity is not necessary for the HA algorithm because the calculation time can be predefined as a stop criterion. The goodness of the solution which depends strongly on the calculation time will be analyzed. In comparison with the other algorithms, $R_{Thres}$ is set to zero.

### 7.1.6 Evaluation and Improvement

Four scheduling methods are used to solve the scheduling problem with the same request matrix **Y** in the same network for comparison. Three network models in Sec. 4.2.2 are taken into account. For evaluating the performances of the algorithms, it is assumed that the channel quality of the connections between each user and its master are the same for all users.

#### 7.1.6.1 Service Model

For evaluation purposes, it is assumed that at the scheduler, a user $U_i$ requests a maximum of $y_{i,j}^{\max}$ instances of a service type $q_j$. As previously mentioned, the duration of each service (e.g. Voice or Video) is assumed to be negative exponentially distributed with a mean value of $E\{T_{Service}\}$ as proposed in [Chan00] and [GuLu07]. Since the scheduling works with the discrete time frame (TF), the geometric distribution is used for modeling the duration of the services. It is also assumed that a service is requested from a user following a Poisson process. However, with our discrete time system, service arrival process is approximated by a Bernoulli process and the probability of requesting a service of a user in a time frame is $p_{on}$. Fig. 7-7 presents a service duration during the run-time.

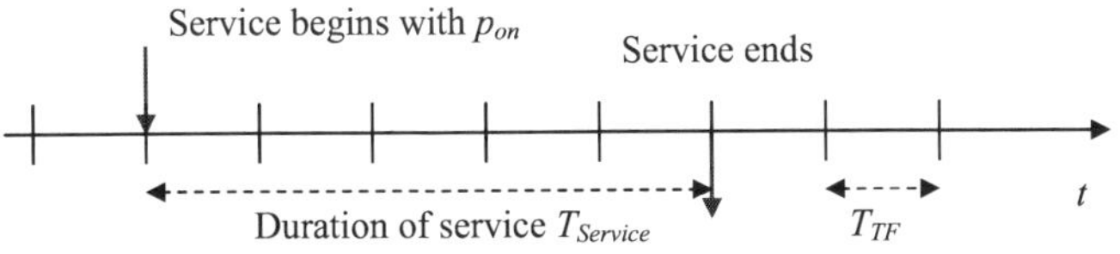


*Figure 7-7: Service process in discrete time*

Let $p_{off}$ be the probability of ending a service in a time frame. Then with the geometric distribution of the service duration, the $p_{off}$ is calculated by:

$$p_{off} = \frac{T_{TF}}{E\{T_{Service}\}} \tag{7-17}$$

The model of such service can be presented by a simple Markov chain presented in Fig. 7-8.

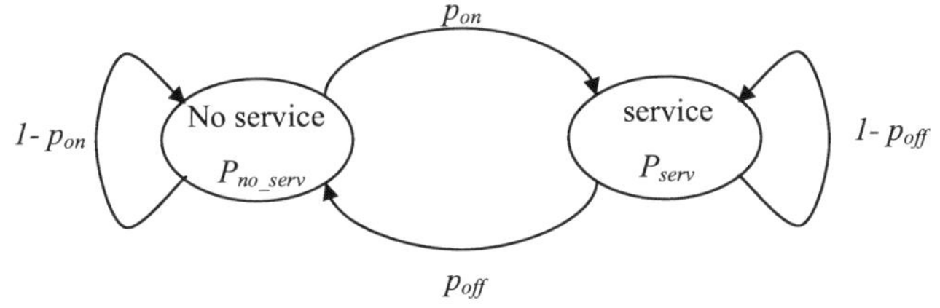


*Figure 7-8: Markov model of a service*

The relationship between the services, users, and scheduling in a base station, which is used in simulation, is presented in Fig. 7-9. Each user $U_i$ consists of $y_{1,1}^{\max}$ service generators for each type of service $q_j$. The service requests from these generators are collected by the *BS* in every time frame for updating the **Y** matrix.

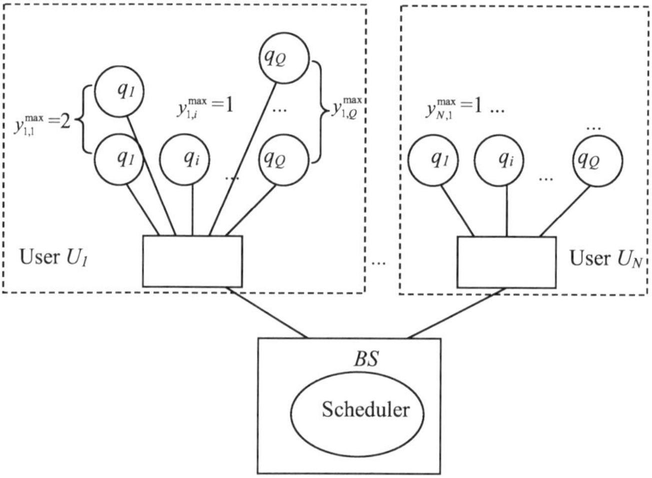


*Figure 7-9: Relationship between services, users, and scheduler in a base station*

#### 7.1.6.2 Data Rate Model of a PLC Link

A simple PLC network using the OPERA specification which consists of one BS and one user has been built up in our laboratory to evaluate the data rate variations during run-time. The PLC modems are the same as described in [D11OP07]. Two PLC links, both with lengths of about 60 - 80 m, were used. Each link consisting of power lines from a power outlet in a room to another outlet in another room was measured for one day. The results have been logged with an integration period of one second. The probability mass function (PMF) of the measured data rate of each link is presented in Fig. 7-10a and Fig. 7-10b.

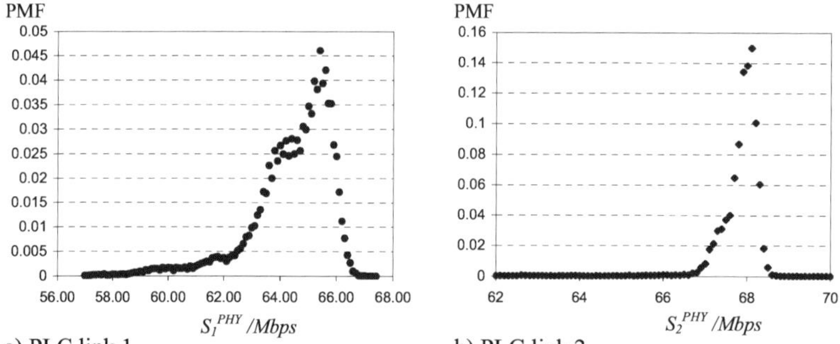


a) PLC link 1

b) PLC link 2

*Figure 7-10: Distributions of measured data rates*

| | mean $\mu$/*Mbps* | variance $\sigma^2$/*(Mbps)*$^2$ |
|---|---|---|
| PLC Link 1 | 64.4 | 2.05 |
| PLC Link 2 | 67.7 | 1.24 |

*Table 7-2: Parameters of measured data rates*

The probability distribution of the data rates may be approximated by a normal distribution, in which the mean and variance are presented in Tab. 7-2. However, in each link, the data rates are not symmetric around the mean value. Therefore, it is assumed that the data rate of each link is a random variable modeled by a Binomial distribution with two parameters $n$ and $p$: $B(n,p)$. The probability of achieving a data rate $k$ with the given maximum data rate $n$ is calculated by the following probability mass function

$$P(k;n,p) = \frac{n!}{k!(n-k)!} p^k (1-p)^{n-k} \tag{7-18}$$

Other PLC links may have different kinds of data rate distributions. In this work, for simulation of the data rate variation, it is assumed that the data rate of each simulative PLC link is also a variable modeled by a Binomial distribution. Using the inversion method in [Lehn09], a distribution table consisting of the data rates in accordance with a Binomial distribution is generated at the beginning of a simulation. After that, only one uniform random variable is needed for updating the data rate of a link. This random number is the position in the distribution table to get the new value of the data rate.

In accordance with the channel measurement results in [D15OP07], the physical channels are stable for one or several seconds. For numerical evaluation, the Binomial generator generates a new data rate with the parameter $p$=0.9, and parameter $n$ calculated by: $n = \mu/p$. The duration between two generations is uniformly distributed between 1 and 20 time frames (between 240 ms and about 5 seconds). Fig. 7-11 presents the PMF function of the data rate generated with the Binomial generator with mean $\mu$= 60 throughput units.

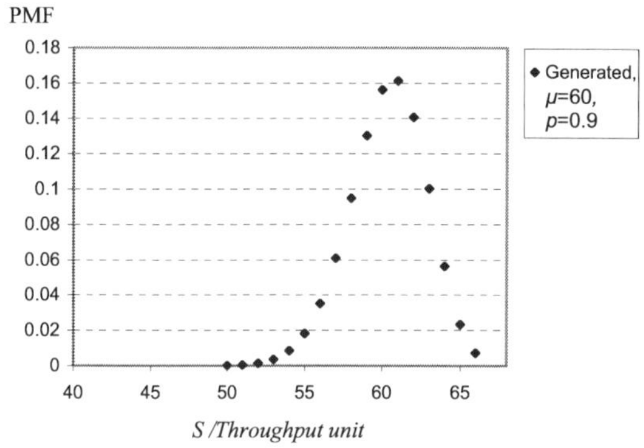


*Figure 7-11: Distribution of generated data rates*

### 7.1.6.3 Evaluation of the Algorithms

The first evaluation is done on the assumption that each user connects to its master with the same average physical-layer data rate. It is also assumed that for each user a fixed amount of resources is reserved for the signaling. The scheduler calculates resources needed for each service with the effective data rate of each user.

For numerical evaluation, the throughput of a VoIP service is normalized to one unit. Each user requests services in four classes with QoS guarantee. For guaranteeing the throughputs, the services $q_1$, $q_2$, $q_3$, $q_4$ request 1, 2, 0.5, 0.125 normalized throughput units (ThU), respectively. In the implementations, if the benefits of giving resources to a new request and to a service previously accepted are equal, the latter one is chosen. Therefore, allocating resources to the service currently guaranteed has higher priority than to a new service request.

| Parameter | Value | Parameter | Throughput |
|---|---|---|---|
| $E\{T_{Service}\}$ | *800TF* | $q_1$ | 1 ThU |
| $p_{on}$ | 1/800 | $q_2$ | 2 ThU |
| | | $q_3$ | 0.5 ThU |
| | | $q_4$ | 0.125 ThU |

*Table 7-3: Service Parameters*

The matrix **Y** is updated in every time frame from each user, in which the parameter $y_{i,j}^{\max}$ is pre-defined. Each service has an average duration of 800 TF and the probability for a service request in a time frame is $p_{on}$=1/800. All the algorithms are studied with the same generated **Y**. The service parameters are summarized in Tab. 7-3. The algorithms are run in three reference networks. For each algorithm, the statistic values are calculated by means of the batch means method [Lehn09]. Except the exhaustive search, each simulation result is conducted by running the algorithms with 10 sub-runs where each lasts for 100.000 time frame duration. The length of each sub-run is set to a high value in order to ensure statistically independency of the simulated samples. The number of sub-runs is not too high to have a not too long simulation time, and according to [Lehn09], too low number of sub-runs results in an inaccurate calculated total variance. The statistical accuracy is set to a confidence probability of 0.95. For comparison, the difference between value $y$ and value $x$ is given by: $d=(y-x)/x\times100$ (in %).

All simulations are run in a computer which has a CPU running at 2.3GHz and sharing 2GB of RAM. In other words, the execution time of an algorithm is measured as a reference for comparing the computational requirements from different algorithms. The heuristic algorithm is executed with 100 iterations per time frame.

For comparison purposes, the networks are configured such that the sum of the requested resources from all users is greater than the available transmission resources (flooded, overloaded system). Tab. 7-4 shows the parameters of the experiments.

| Network | Network $C_1$ | Network $C_2$ | Network $C_3$ |
|---|---|---|---|
| *Size N* | 8 | 25 | 120 |
| $y_{i,j}^{\max}$ | $U_1..U_2=1$<br>$U_3..U_5=2$<br>$U_6..U_8=3$ | $U_1..U_8=1$<br>$U_9..U_{16}=2$<br>$U_{17}..U_{25}=3$ | $U_1..U_{40}=1$<br>$U_{41}..U_{80}=2$<br>$U_{81}..U_{120}=3$ |
| $Q$ | 4 | 4 | 4 |
| $E[S_{U_i}^{E(PHY)}]$ /ThU | 20 | 60 | 240 |

*Table 7-4: Network Parameters*

### *7.1.6.3.1 Comparison of Algorithms*

Fig. 7-12 shows the comparison of the execution time and the fitness achieved by applying different algorithms with the small network $C_1$. Due to the complexity of the exhaustive search mechanism, a sub-run consists of just 100 time frames. As expected, the exhaustive search finds the highest fitness (average 202.2BU) but it takes the longest time to find the solution (average 128.7s). The HA algorithm finds a slightly lower fitness (194.6BU (confidence interval (CI) is [189.5, 199.7])), but within a significantly shorter calculation time (0.85 [0.81, 0.88] ms) than the ES algorithm. The BBS runs longer than the RRS but achieves a better utilization (188.0[183.3, 192.8] BU) compared to 161.1 [156.7, 165.5] BU).

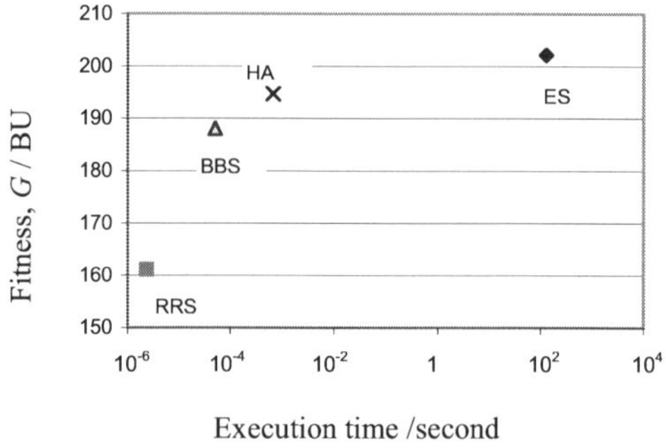


*Figure 7-12: Utilization and execution time with different algorithms for network $C_1$.*

Due to the complexity of the exhaustive search, it is only suitable for small sized networks. If each user requests on average one service for each type, then each service is in one of the two states in a time frame: there is a request or not. In accordance with the time complexity of the ES algorithm presented in Sec. 7.1.4.1, the calculation time for the network $C_2$ ($N$=25) is $2^{(25-8)}=2^{17}$ times the calculation time for the network $C_1$ ($N$=8). No result has been obtained for the ES algorithm for the larger networks $C_2$ and $C_3$ since the simulation takes so long. Therefore, only HA, BBS, and RRS are simulated for the networks $C_2$ and $C_3$. Fig. 7-13 presents the comparisons of the performances of the algorithms with the three reference networks. In the graphs, the confidence intervals are too small to be visible. Detailed numerical results are presented in Appendix A.3 (Tab. A.3 to Tab. A.5).

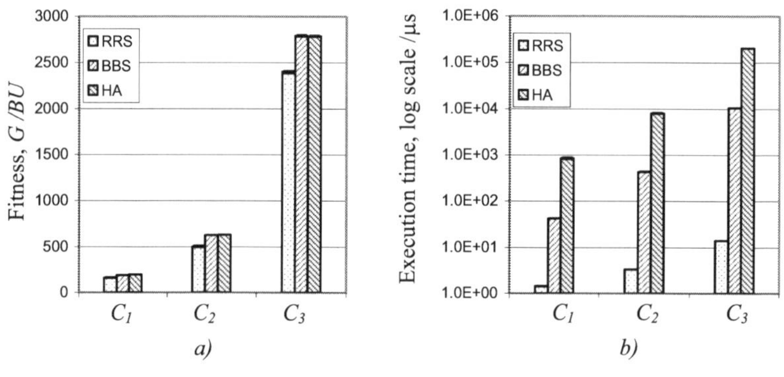


*Figure 7-13: Comparisons of a) fitness and b) execution time with different algorithms for three networks.*

It can be seen from the comparisons that the BBS and the HA algorithm are mostly suitable for solving the formulated optimization problem. For each network, the execution time is exponentially proportional to the network size. Utilizing the BBS, the scheduler can find a good solution (e.g. for $C_3$ as presented in Tab. 7-5, the average fitness is 2791.8 [2775.1, 2808.5] BU) compared to 2402.3 [2385.5, 2419.1] BU) in a relatively short calculation time (10.55[10.50, 10.61] ms). The results from the HA are slightly better than the BBS in the small ($C_1$) and medium ($C_2$) sized networks, and slightly worse than the BBS in the large ($C_3$) sized network. The execution time is, however, significantly longer, e.g. for $C_3$, the HA algorithm needs 205.69 [204.82, 206.56] ms.

| | RRS (Mean [CI]) | BBS (Mean [CI]) | HA (Mean [CI]) |
|---|---|---|---|
| Fitness $G$ / BU | 2402.3 [2385.5, 2419.1] | 2791.8 [2775.2, 2808.5] | 2788.0 [2771.8, 2804.2] |
| $B_U$ /BU | 1277.6 [1269.0, 1286.2] | 1480.8 [1472.6, 1489.1] | 1420.2 [1412.4, 1428.0] |
| $F_U$ | 0.88 [0.879, 0.881] | 0.885 [0.884, 0.887] | 0.920 [0.919, 0.921] |
| Execution time /ms | 14.19 [14.16, 14.23] ($\times 10^{-3}$) | 10.55 [10.50, 10.61] | 205.69 [204.82, 206.56] |

*Table 7-5: Simulation results using different algorithms with PLC Cell $C_3$*

In all three cell sizes, the RRS provides the worst fairness, the fairness of the BBS is higher, and the HA provides the highest fairness. For the HA algorithm, the quality of the result depends on the limitation of the duration for the search, reflected by the number of iterations. For all three networks, the fitness achieved from the BBS and HA increases about 25% to 30%, compared with the fitness from the RRS. One improvement to shorten the calculation times, based on the convergence characteristic of the heuristic algorithm, is presented in the following section.

#### *7.1.6.3.2 Resource Requests Analysis*

The relationship between the number of requested and the number of granted services is analyzed. To do so, the users generate different service requests in every time frame. The parameters in Tab. 7-4 are used. Both the BBS and HA algorithms produce similar results. Fig. 7-14 shows the comparison of these numbers of services for each user when utilizing the HA algorithm. The results are presented for network $C_2$ with the parameters given in Tab. 7-4. The algorithm tends to allocate more resources to the users with a higher number of requested services than the users with lower requests. For each user in the high loaded network, the number of unguaranteed services is almost proportional to the number of requested services.

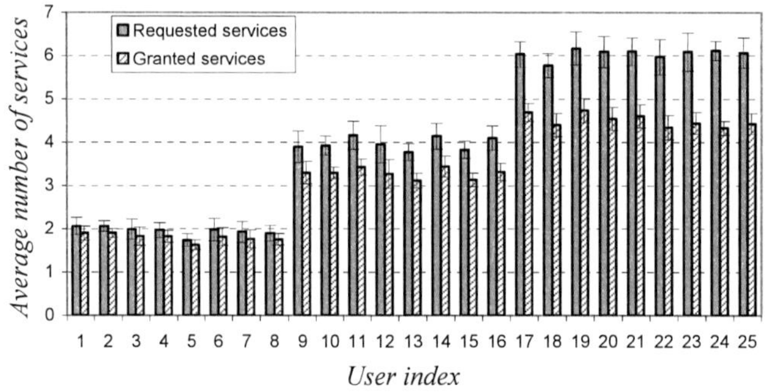


*Figure 7-14: Comparison of the number of request services and granted service for each user in network $C_2$ with HA, with the same channel qualities*

### *7.1.6.3.3 Channel Quality Analysis*

In order to analyze the effect of the channel qualities of different users to the grant matrix, each user is assumed to request a fixed and the same number of requests. The results from the BBS and HA algorithm are similar when channel qualities of the users to its master are different. The HA algorithm running with the PLC cell $C_2$ is taken for analysis. Assuming that the users from $U_1$ to $U_8$ (UG1) have the highest effective data rate, from $U_9$ to $U_{16}$ (UG2) have a lower data rate, and from $U_{17}$ to $U_{25}$ (UG3) have the lowest data rate, the average data rates in the groups of users are set as follows:

$$E[S_{U_{G1}}^{E(PHY)}] = 2E[S_{U_{G2}}^{E(PHY)}] = 3E[S_{U_{G3}}^{E(PHY)}] \tag{7-19}$$

With the average effective data rates, to guarantee a VoIP with one throughput unit, users in UG2 need one resource unit, while users in UG1 need half of a resource unit and users in UG3 need 3/2 resource units. Network parameters are presented in Tab. 7-6.

| | $U_1$- $U_8$ (group UG1) | $U_9$- $U_{16}$ (group UG2) | $U_{17}$- $U_{25}$ (group UG3) |
|---|---|---|---|
| $y_{ij}^{\max}$ | 2 | 2 | 2 |
| $Q$ | 4 | 4 | 4 |
| $E[S_{U_i}^{E(PHY)}]$ /ThU | 120 | 60 | 40 |

*Table 7-6: Network Parameters*

Fig. 7-15 presents the relationships between the number of granted services and request services of different users. Because the HA algorithm tries to maximize the benefit, it tends to assign more resources to the users who can transmit with a higher data rate. However, for keeping the fairness between the users, the users with a lower data rate are also granted resources. The ratio of the number of granted and requested services is higher than the ratio of the transmission data rates between the users who have different channel qualities. The RRS algorithms, however, tend to uniformly distribute transmission resources to the users. Therefore, the number of granted resources is proportional to the data rate of the users. The fairness between the users is low, as shown in Fig. 7-16.

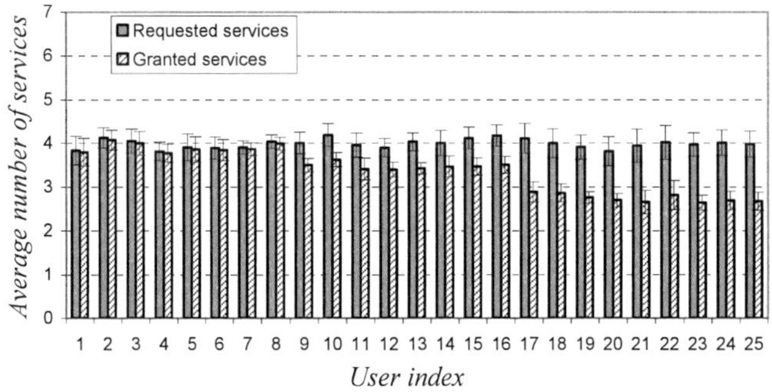


*Figure 7-15: HA algorithm: comparisons of the number of request services and granted service for each user in network $C_2$ with different channel qualities*

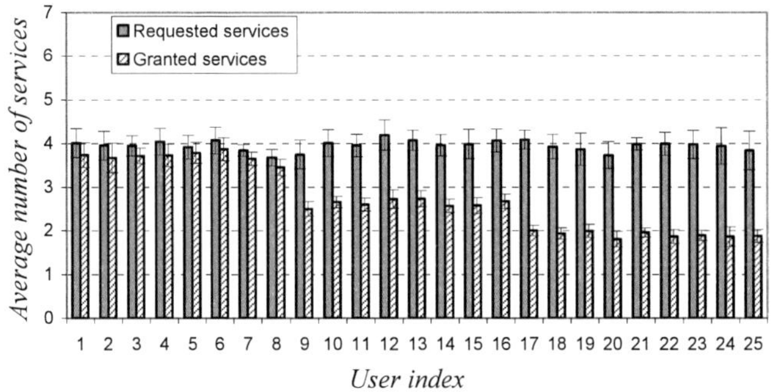


*Figure 7-16: RRS algorithm: comparisons of the number of request services and granted service for each user in network $C_2$ with different channel qualities*

#### 7.1.6.4 Improvement of the HA with an Adaptation Strategy (HAAS)

To shorten the calculation time when utilizing the HA algorithm, the so-called adaptation strategy (HAAS) is proposed. Let us run the HA to solve the optimization problem with the network $C_2$ with a long execution time. The convergence characteristic of the algorithms is presented in Fig. 7-18 in which the fitness $G$ is presented during the first 1000 iterations. The utilization increases quickly at the beginning of the search and slower when the fitness is near the optimum value. The fitness found in an iteration depends on the fitness in the previous one, thus the result from the HA algorithm depends on the initial solution.

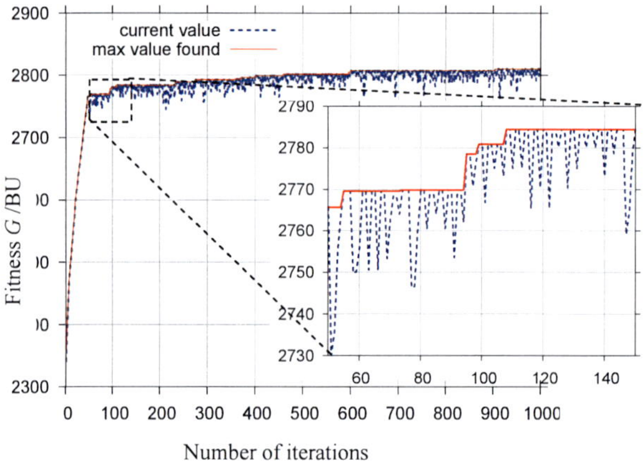


*Figure 7-17: Convergence characteristic of the HA algorithm*

When considering only one service, its average duration is relatively long compared to a time frame, e.g. average duration of a VoIP session is 3 minutes compared to $T_{TF}$=240 ms. Therefore, the probability that the status of a service changes after a time frame is relatively small. The scheduler has to perform the scheduling algorithm in every time frame. This means that the probability of the changes in the matrix **Y** from a time frame to the next time frame in which the scheduler re-executed is also small. The accurate parameters are given in the analysis sections. The strategy is to adapt the solution in the previous time frame with the new request matrix **Y** to create a better initial solution before running the HA algorithm. The adaptation makes the initial solution feasible by changing the grant matrix **X** to meet the constraints in (7-5) and (7-12). The flowchart and position of the adaptation procedure in the scheduling module are presented in Fig. 7-18.

With a lower probability of the variation of the matrix **Y**, the adaptation procedure produces an initial solution closer to the optimum solution. If the matrix **Y** is not changed after a time frame, the HA algorithm runs further as if it is allowed to run with twice the number of iterations. Therefore, it is expected that the HAAS algorithm finds a better solution, or needs a shorter execution time compared to the HA algorithm.

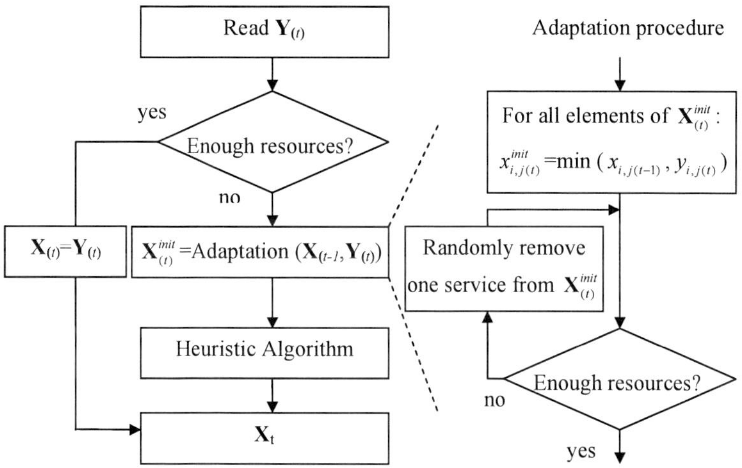


*Figure 7-18: Adaptation procedure and its position in the scheduling module*

For comparing the performance, both HA and HAAS are executed with the same request matrix **Y** and network $C_3$. Fig. 7-19 presents the fitness produced by both algorithms during 10 time frames in two cases: the number of one execution in a time frame (TF) is 100 iterations and 10 iterations. Conducted in a long simulation run, the fitness and execution time are presented in Tab. 7-7. Detailed numerical results are given in Appendix A.3 (Tab. A.6).

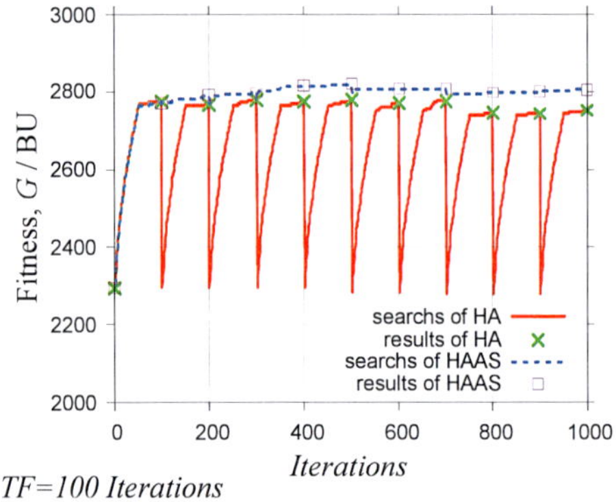


*a) 1TF=100 Iterations*

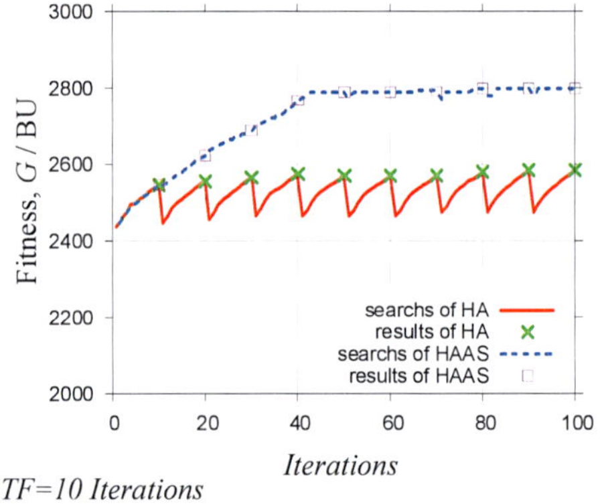


*b) 1TF=10 Iterations*

*Figure 7-19: HA and HAAS algorithms comparisons a) 1 TF=100 iterations, b) 1TF=10 iterations*

When running the HAAS algorithm with 100 iterations in a time frame, the fitness is slightly improved (2.20%), but the execution time is significantly shorter (-42.20%). It is because the HA has enough time to reach a fitness near the optimum value. The adaptation scheme can only shorten the search time. If the HA is run for only 10 iterations, the result from the algorithm has not yet reached the local optimum. The result is still in the improvement phase from the initial solution (RRS). On the other hand, the HAAS result from the previous time frame is used for the next run, and with the adaptation procedure, the initial solution is already higher than the RSS. HAAS therefore can improve the search results, e.g. the fitness improves by about 12%.

Comparing between the HAAS running with 10 iterations per time frame and the HA running win 100 iterations per time frame, not only the fitness from HAAS is slightly higher than that from the HA (average: 2828BU compared to 2787BU), but the execution time is more than 17 times shorter (average: 12.4 ms compared to 205 ms).

| Algorithm | Parameter | Mean | Confidence Interval |
|---|---|---|---|
| 100 iterations per TF | | | |
| HA | Fitness *G* / BU | 2788.0 | [2771.8, 2804.1] |
| | Execution time /ms | 205.7 | [204.8, 206.6] |
| HAAS | Fitness *G* / BU | 2849.3 | [2838.7, 2860.0] |
| | Execution time /ms | 118.9 | [116.2, 121.5] |
| 10 iterations per TF | | | |
| HA | Fitness *G* / BU | 2523.5 | [2507.9, 2539.0] |
| | Execution time /ms | 11.7 | [11.6, 11.8] |
| HAAS | Fitness *G* / BU | 2828.3 | [2811.2, 2845.3] |
| | Execution time /ms | 12.5 | [12.3, 12.6] |

*Table 7-7: Simulation results with HA and HAAS algorithms*

The improvement by using the HAAS however depends strongly on the results from the adaptation of the initial solution. Its performance becomes worse with more frequent variations (service requests, physical channel varies) in the network in every time frame. Therefore, the HAAS algorithm is suitable for the network in which variations in the network after each time frame are rare. In the worst case, its performance will fall back to the results from the HA algorithm.

#### 7.1.6.5 Conclusions

The BBS and HA are most suitable for solving the scheduling problem. Due to the long execution time of the HA algorithm in a large PLC network, an adaptation strategy is proposed to form the HAAS algorithm. Compared with the HA algorithm, if the variations in the network are rare, this algorithm produces better results within a relatively short calculation time. The BBS requires simple implementation but performs slightly better than the HA and worse than the HAAS when the network is large (as in case of the network $C_3$).

## 7.2 Utilization of a TDMA/OFDMA Scheme

By utilizing the diversity of the frequency dimension with the OFDM modulation scheme, the transmission between two PLC nodes can be individually organized in different frequency channels. When a user can use different channels (OFDMA) for its transmission in a time period (TDMA), the resources are presented in two dimensions: frequency and time

(TDMA/OFDMA). Each resource unit, formed by a frequency channel and an OFDM symbol, can be used to carry the data of a specific service for a particular user.

As stated by [AyyaCh05] and [KwonSe09] for other systems, the improvement of transmission capacities by utilizing the TDMA/OFDMA results in higher costs of computation and complexity, as compared to the TDMA. The principle of the scheduling is that a user $U_i$ is scheduled to transmit a longer duration in higher data rate channels. Another requirement is that the MAC and PHY layer have to be modified to be able to address two dimensional resource units.

When each channel can be scheduled and used to transmit data independently, each channel in the TDMA/OFDMA scheme is scheduled with the TDMA basis. The uplink and downlink transmissions are organized in the same channel with different time periods. The task of the scheduling module is to decide the starting point and duration of each transmission in each channel for each node. In accordance with these transmission durations, the transmitter splits data and transmits it in different channels. On the receiver side, data transmitted from different channels is re-combined to form the original data before sending to the upper layer.

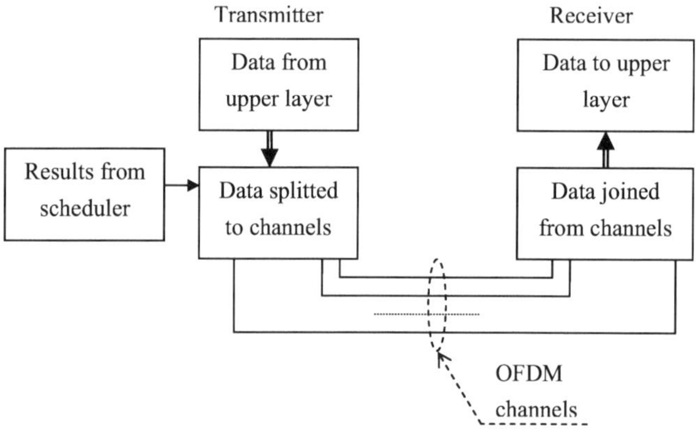


*Figure 7-20: Transmission principle with multiple frequency channels.*

The decision of the scheduler, including the transmission time for each user in each channel, is investigated. Since organizing transmissions in the uplink and the downlink are similar, the scheduling is investigated for downlink traffic. The problem formulation is also extended from the TDMA problem formulation presented previously.

### 7.2.1 Model of Transmission Data Rate in Channels

Assuming that the PLC cell allocated $B_C$ channels indexing from $f_1$ to $f_{Bc}$, i.e. they are the results from the DICRA protocol presented in chapter 6, each channel is a group of contiguous OFDM carriers. For a user $U_i$ , the data rate in a channel $f$ is $S_{U_1}^{PHY,f}$ . The transmission data rate between a user and its master is the sum of the transmission data rates in all available channels.

$$S_{U_i}^{PHY} = \sum_{f=1}^{B_C} S_{U_i}^{PHY,f} \tag{7-20}$$

Based on the measurement results presented in [D15OP07], the amount of data that an OFDM carrier can transmit depends on the positions of the users, and the frequency of the carrier. It is varied during run-time. To evaluate the scheduling method, three types of the transmission data rates distributed over the channels are assumed.

Fig. 7-21 demonstrates three types of data rates depending on the channel index. In type 1 ($CHT1$), the average data rate is uniformly distributed over the channels. This type may be observed in a link with a small frequency spectrum and with low disturbances. In this type, the relationship between the total data rate and the average transmission data rate in a channel is

$$E[S_{U_i,CHT1}^{PHY,f}] = \frac{S_{U_i}^{PHY}}{B_C} \tag{7-21}$$

In type 2($CHT2$), the average transmission data rate linearly decreases when the carrier frequency increases. This type may be observed in links using a wide frequency spectrum. The average data rate is formulated by

$$E[S_{U_i,CHT2}^{PHY,f}] = \frac{(B_C - f) \cdot S_{U_i}^{PHY}}{B_C^{\ 2}} + \frac{S_{U_i}^{PHY}}{2B_C} \tag{7-22}$$

The average data rate linearly increases when the frequency increases in type 3 ($CHT3$). This type may be observed when there are more intensive disturbances in the low frequency band. The average transmission data rate is calculated by

$$E[S_{U_i,CHT3}^{PHY,f}] = \frac{(f - 1) \cdot S_{U_i}^{PHY}}{B_C^{\ 2}} + \frac{S_{U_i}^{PHY}}{2B_C} \tag{7-23}$$

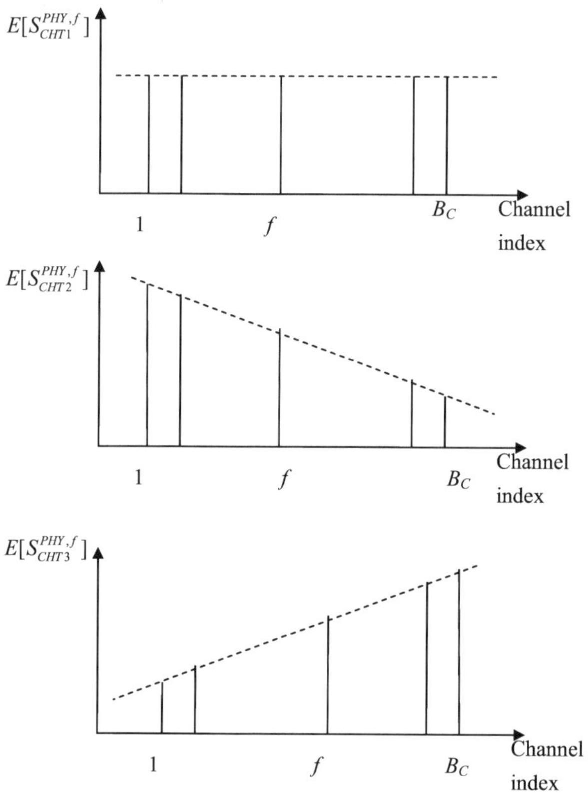


*Figure 7-21: Three models of the data rates over the channels*

### 7.2.2 Scheduling with TDMA/ OFDMA

For the formulation of the problem, the transmission time ($t \in [1; T_{TF}]$)) in a time frame for each user ($U_i$, $i \in [1;N]$) in a channel ($f \in [1;B_C]$) is necessary. The granularity of the time dimension is the number of OFDM symbols. The transmission durations of the users in the channels are represented in the allocation matrix $\mathbf{T}=\{T_{i,f}\}$. Fig. 7-22 presents an allocation of the transmission durations for different users in different channels.

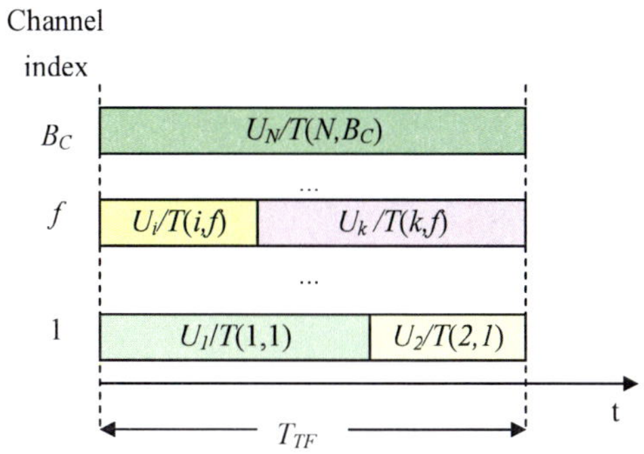


*Figure 7-22: Organization of transmissions in a time frame*

Similar to (7-3) the throughput of the user $U_i$ to transmit a service $q_j$ in a channel $f$ is calculated by the transmission duration $T_{i,j,f}$ in this channel

$$\beta_{i,j}^{f} = S_{U_i}^{PHY,f} \cdot \frac{T_{i,j,f}}{T_{TF}} \tag{7-24}$$

Therefore, to guarantee a service which requires throughput $\beta_{i,j}$, the throughput of the user to transmit in all channels is

$$\sum_{f=1}^{B_C} \beta_{i,j}^{f} = \sum_{f=1}^{B_C} S_{U_i}^{PHY,f} \cdot \frac{T_{i,j,f}}{T_{TF}} \geq \beta_{i,j} \tag{7-25}$$

Each service $q_j$ of a user $U_i$ consumes an amount of resources:

$$T_{i,j} = \sum_{f=1}^{B_C} T_{i,j,f} \tag{7-26}$$

Similar to the TDMA scheme, an overhead is needed for organizing the transmission in each channel. We assume that the overhead is the same for all channels and the same with the TDMA scheme (user $U_i$ requires $T_{U_i}^{OH}$ for organizing the transmission in a channel). The constraints of the transmission durations in all channels are

$$\forall f \in [1, B_C], T_{i,f} > 0: \ \sum_{i=1}^{N} (T_{i,f} + T_{U_i}^{OH}) \leq T_{TF} \tag{7-27}$$

The optimization problem can be formulated as: to find a transmission allocation matrix **T** and the grant matrix **X** in order to

*maximize* the fitness $G$ in (7-14)

such that

(7-5), (7-25), and (7-27) are fulfilled.

### 7.2.3 Algorithm and Evaluation

Neglecting the overhead for addressing the resources, the transmission resources for transmitting payload are the same as in the case of using the TDMA scheme. Solving this problem with HAAS requires a long execution time, e.g. several seconds for the PLC cell $C_2$ and $C_3$, which is not feasible to be implemented in a BS. Therefore, one extension of the BBS is proposed, called EBBS algorithm. Compared to the BBS, the EBBS algorithm requires one additional procedure to organize the data into different channels. It is done by allocating the resources to the service which results in the highest ratio of the fitness per total transmission durations in different channels. Let $\Delta G^{+}_{i,j}$ be an additional fitness gained by assigning resources to service $q_j$ of a user $U_i$: $\Delta G^{+}_{i,j} = G(x_{i,j}+1) - G(x_{i,j})$. The algorithm contains the following steps:

- <u>*Step 1*</u>: Calculate $\frac{\Delta G^{+}_{i,j}}{T_{i,j}}$ for all possible assignments for all users and all services. The

  calculation for a value of $\frac{\Delta G^{+}_{i,j}}{T_{i,j}}$, for user $U_i$, service $q_j$

  - <u>*Step 1.1*</u>: search for the channel $f$ which has available transmission resources and with highest data rate.
  - <u>*Step 1.2*</u>: Calculate transmission duration for service $q_j$ in that channel $f$
  - <u>*Step 1.3*</u>: If available transmission resources in $f$ are greater or equal to the resources needed, the service is allocated entirely in channel $f$. Otherwise, available resources in that channel are allocated to carry part of the service. Repeat S*tep 1.1* to *1.3* until the rest of the service is allocated or there are no more available resources. The resources $T_{i,j}$ is calculated by (7-26).

- <u>*Step 2*</u>: Assign resources $T_{i,j}$ to the request which achieves the highest $\frac{\Delta G^{+}_{i,j}}{T_{i,j}}$

- <u>*Step 3*</u>: Repeat the Step 1 and Step 2 until it meets one of the following conditions: all service requests are granted or there are no more resources in all channels.

To evaluate the performance of the EBBS algorithm, the users are divided into three groups with three transmission data rate models. The data rate in each channel is again generated in accordance with the Binomial distribution. For numerical evaluation, the Binomial random generator generates a new data rate with the parameter $p$=0.9 and the given mean data rate. Again, each simulated value is calculated after the simulation runs with 10 sub-runs, each lasts for 100.000 time frames.

#### 7.2.3.1 Channel Occupation

For analyzing the schedule, the small network $C_1$ is considered. For comparing the channel occupations between the users with different channels, it is assumed that all users request the same number of services. The model of the channel quality of users $U_1$ and $U_2$ is *CHT*1, of the users $U_3$ to $U_5$ is *CHT*2, $U_6$ to $U_8$ is *CHT*3. For numerical evaluation, the average transmission data rates of different users in different channels, which are represented by the throughput units (ThU), are given in Tab. 7-8 . These values are used to generate the data rate of each user in each channel during run-time.

| $E[S_{U_i}^{PHY,f}]$/ThU / Channel index, $f$ | $U_1$, $U_2$ (group UG1) | $U_3$, $U_4$, $U_5$ (group UG2) | $U_6$, $U_7$, $U_8$ (group UG3) |
|---|---|---|---|
| 1 | 3.4 | 4.5 | 1.7 |
| 2 | 3.4 | 3.9 | 2.2 |
| 3 | 3.4 | 3.4 | 2.8 |
| 4 | 3.4 | 2.8 | 3.4 |
| 5 | 3.4 | 2.2 | 3.9 |
| 6 | 3.4 | 1.7 | 4.5 |

*Table 7-8: Average transmission data rate in different channels*

One transmission allocation matrix achieved from simulation is presented in Fig. 7-23. The users $U_3$, $U_4$, $U_5$ transmit with the highest data rate in channels $f_1$ and $f_2$, therefore their transmissions are mainly allocated in these channels. Similar to that, users $U_6$, $U_7$, $U_8$ transmit with the highest data rate in channels $f_5$ and $f_6$, so they are mostly transmitted in these channels. Users $U_1$ and $U_2$ transmit with the same data rate in all channels, so their transmission durations distribute over the rest of the channels.

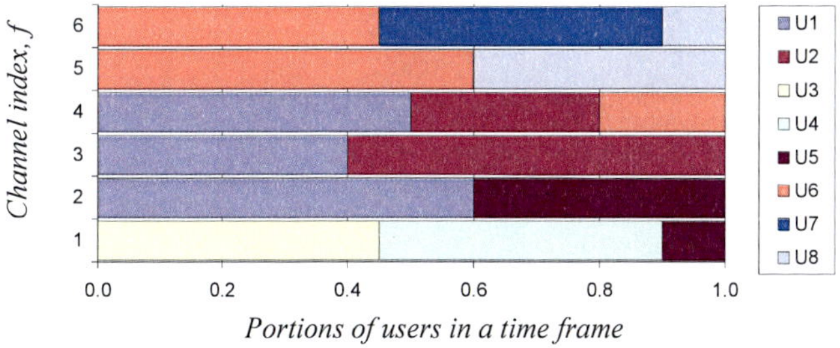


*Figure 7-23: Transmission durations of different users in different channels, network $C_1$ with EBBS*

The simulation is also run with network $C_2$ (with 20 channels) and $C_3$ (with 80 channels). In each network, the users are also divided into three groups and each group has one channel quality model. One allocation matrix for the network $C_2$ is presented in Fig. 7-24. The results are similar to the case of $C_1$. Users in group UG2 are mostly scheduled in the channels in type CHT2, users in group UG2 are scheduled in the channels in type CHT3, and the users in group 1 are scheduled in the rests of the channels. The results prove the tendency that each user is allocated to transmit in the channels that have higher data rates than the other channels.

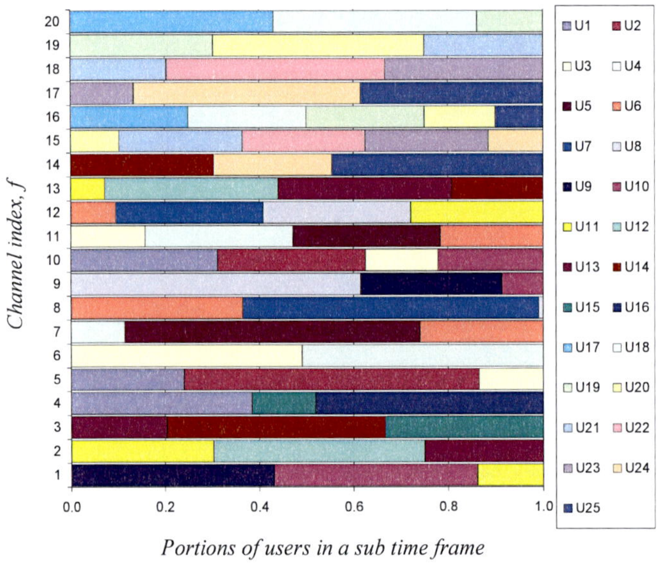


*Figure 7-24: Transmission durations of different users in different channels, network $C_2$ with EBBS*

#### 7.2.3.2 Performance Evaluation

For comparison between the TDMA and TDMA/OFDMA schemes in each time frame, the total data rates of all channels are used for the TDMA scheme. In each time frame, the same request matrix **Y** is generated and used to run all the scheduling algorithms. Fig. 7-25 presents the fitness comparison of the BBS and HA with TDMA and the EBBS with the TDMA/OFDMA scheme. Some confidence intervals are too small to be seen.

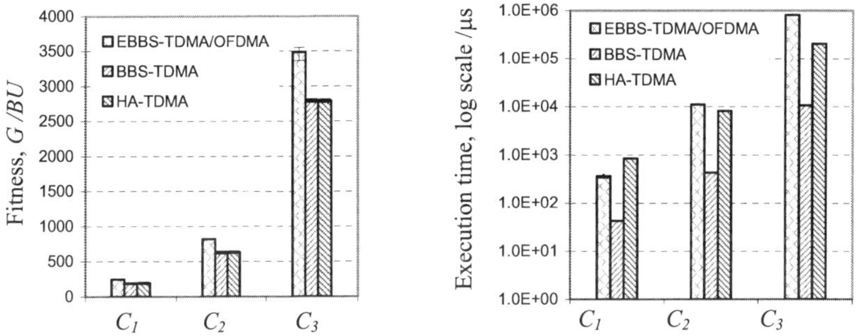


*Figure 7-25: TDMA/OFDMA and TDMA comparisons of a) fitness and b) execution time with different algorithms for three networks.*

It can be seen from the simulation that the TDMA/OFDMA scheme results in a better fitness. The fitness of a scheduling solution reflects the resource utilization. Better resource utilization means that more services are possible and the PLC cell can transmit with higher transmission capacity compared to the TDMA scheme.

| | BBS-TDMA | | EBBS-TDMA/OFDMA | | |
|---|---|---|---|---|---|
| | Mean | Confidence Interval | Mean | Confidence interval | Different /% |
| Fitness $G$ / BU | 188.0 | [183.2, 192.8] | 246.2 | [237.9, 254.5] | 30.9 |
| $B_U$ /BU | 99.3 | [96.8, 101.9] | 124.9 | [120.8, 129.0] | 25.7 |
| $F_U$ | 0.892 | [0.887, 0.897] | 0.970 | [0.966, 0.973] | 8.7 |
| Execution time /ms | 0.042 | [0.041, 0.043] | 0.365 | [0.345, 0.384] | 764.7 |

*Table 7-9: BBS with TDMA and EBBS with TDMA/OFDMA with PLC Cell $C_1$*

The fitness improvement by utilizing the EBBS compared to the BBS in three networks is presented in Appendix A.3 (Tab. A.7). For the cell $C_1$ as presented in Tab. 7-9, the fitness improvement is about 30.92% ($G^{EBBS}$=246.2 [237.9, 254.5] BU, compared to $G^{BBS}$= 188.0[183.3, 192.8] BU), but it requires about 7 times longer calculation time. The improvement is achieved by both improvement of benefit and fairness. For the larger network $C_2$ and $C_3$, the fitness increases about 30.36% and 24.89%, respectively. However, with larger networks, the execution time of the EBBS is many times longer. In the case of the largest network $C_3$, the average execution time is 815ms.

With TDMA/OFDMA, the users are scheduled to transmit in the channels with high data rates compared to the other users. In all three networks, the combination of the TDMA/OFDMA provides an improvement of resource utilization compared to the TDMA scheme. The improvement depends on the variation of the data rates between the channels. With one additional frequency dimension, the EBBS has a higher computation complexity thus requiring a longer execution time.

## 7.3 Summary

In this chapter, scheduling algorithms for the intra-PLC-cell are investigated. The service admission control is the key for deciding if a service is guaranteed or not. With the assumption of using the available MAC and PHY layers from the OPERA specification, the problem is to schedule in a pure TDMA system. By formulating the transmission resource allocation problem into an optimization problem, the problem can be solved by four algorithms, namely ES, RRS, BBS, and HA. Their performance parameters (fitness of the solution and execution time) depend strongly on the size of the network. Each network provides four types of services that require QoS guarantee.

Generally, the improvement in resource utilization results in the higher computation complexity and thus a longer execution time. A reduction of the execution time for the HA algorithm is achieved by means of an adaptation strategy, called HAAS algorithm. The improvement of the HAAS compared to the HA depends on the variations in the network during run-time, i.e. resource requests and channel data rates. The simple BBS algorithm allocates the resources in a relatively short execution time but with almost the same fitness compared to the HA. Both BBS and HAAS are suggested for scheduling in the PLC cell with the TDMA access scheme. Their performances outperformed the RRS algorithm.

Additionally utilizing the OFDMA scheme, the search space is larger due to the additional frequency dimension. In this case, each frequency channel can be independently scheduled. The data from users are split to transmit in different channels, based on the channels' data rates. By running our proposed EBBS algorithm, the resource utilization can be increased. The improvement of TDMA/OFDMA depends on the differences in the channels' data rates. However, it requires complex computation which may not be feasible for use within large PLC networks.

# Chapter 8 Conclusions and Outlook

This thesis investigates the resource allocation in multi-cell PLC networks. The proposed resource allocation scheme maximizes the utilization of the limited transmission resources, while providing QoS guarantee and fairness between users. The main contributions of this thesis are as follows:

- By utilizing a cellular concept, the so-called hierarchical resource allocation scheme is proposed. The frequency spectrum is divided into smaller bands, called channels. Each channel consists of several consecutive OFDM carriers. In the inter-PLC-cell level, the channels are allocated to the PLC cells. In the intra-PLC-cell level, the allocated channels are shared by communication services from the users.
- For inter-PLC-cell resource allocation, the so-called Distributed Inter-Cell Resource Allocation (DICRA) protocol is proposed. The protocol requires regular exchanges of channel information between neighboring PLC cells by means of the so-called channel announcement messages (CA-Msg).
  - Two procedures, namely "Seize" and "Release", are proposed. The protocol dynamically reallocates the channels between the PLC cells in accordance with the changes of the traffic demands from the cells.
  - The protocol works in a distributed fashion, in which each BS in a PLC cell locally determines its allocated channels based on channel information from its neighbors.
  - The protocol maintains fairness between users in different cells in terms of the number of channels per active user, achieves a high channel reuse factor, and avoids usage of the same channels in neighboring cells.
  - The performance of the protocol strongly depends on the number of neighbors of a cell, but not on the size of the PLC network.
  - The proposed DICRA protocol can be applied to access- and inhome PLC systems and coexists between them. It can also be applied to other technologies that support multi-cell structures.
- On the intra-PLC-cell level, the resource allocation problem is formulated as an optimization problem. It is assumed that the MAC and PHY layers from the OPERA specification are used. Two objectives, the benefit of resource utilization and the fairness between users, are combined into a fitness function.

- o Fitness and execution times of the exhaustive search (ES), round robin scheduling (RRS), simulated annealing based heuristic algorithm (HA), and of the proposed benefit based selection (BBS) algorithm are evaluated.
- o Both the BBS and HA algorithms are suggested for the scheduling task. Despite requiring longer execution times, they perform significantly better than the RRS in terms of resource utilization, number of guaranteed services, and fairness between users. The HA requires a very long execution time with a large PLC cell.
- o The adaptation scheme is proposed to speed up the heuristic algorithm (called HAAS). The performance of the HAAS depends on the variations during run-time of traffic demands and channel qualities in the PLC cell.
- o The scheduling problem is formulated for the combination of OFDMA and TDMA. An extension of BBS (called EBBS) is proposed to solve this problem.
- o Network utilization using the OFDMA/TDMA is higher than with the TDMA when the channels have different data rates. E.g. with the investigated networks and channel models, the fitness improvements are about 25-30%. The improvement is achieved at the cost of computation time.

The results achieved in this thesis can be further enhanced. Following ideas can be considered in further work:

- The DICRA protocol is proposed to be used in different situations of the multi-cell PLC network. In a concrete situation, the fairness definition in use may not fulfil requirements from the network owner, e.g. fairness between an access cell and an inhome cell may differ from the fairness between two access cells. Therefore, it can be further designed to operate in several modes, each used in a specific situation.
- In a multi-cell PLC network operated by one service provider, a protocol that supports a user to select one of the nearby BSs may improve the connected channel quality. It can also allow dynamic optimization of the PLC cell structure.
- Since the OFDMA/TDMA scheme is solved based on several assumptions, improvement is necessary. Firstly, the PHY and MAC layers have to be designed to support the OFDMA access scheme. Secondly, due to the calculation complexity of the problem, a distributed approach can be applied, e.g. using slave scheduling modules in the repeaters and users. The algorithm could also be optimized to execute in less time, e.g. by grouping channels and symbols into blocks or by running several parallel modules in additional co-processors.

# References

[AbadTo05] J. Abad, L. M. Torres, J. C. Riveiro, "OFDM and Wavelets Performance Comparison in Power Line Channels", *Proc. of 9$^{th}$ ISPLC 2005*, pp. 341 – 345, Vancouver, Canada, Apr. 6-8, 2005.

[Alfy10] E. –S. M. El-Alfy, "Chapter 13 Routing and Traffic management", *Selected Topics in Communication Networks and Distributed Systems,* World scientific publishing Co. Ltd. ISBN-13 978-981-283-943-5, pp. 343-400, 2010.

[AmalCa03] E. Amaldi, A. Capone, F. Malucelli, "*Planning UMTS Base Station Location: Optimization Models with Power Control and Algorithms",* IEEE Trans. on Wireless Communications, vol. 2, iss. 5, pp. 939 – 952, 2003.

[AnanSr03] S. Anand, A. Sridharan, K.N. Sivarajan, "Performance analysis of channelized cellular systems with dynamic channel allocation", *Vehicular Technology, IEEE Trans*. vol. 52, iss. 4, pp. 847 – 859, July 2003.

[AndFr95] A. Baiocchi, F. D. Priscoli, F. Grilli, F. Sestini. "The Geometric Dynamic Channel Allocation as a Practical Strategy in Mobile Networks with Bursty User Mobility," *IEEE Transactions on Vehicular Technology*, vol. 44, iss. 1, pp. 14-23, Feb. 1995.

[AntoSt97] Z. Antoniou, I. Stavrakakis, "Earliest due first scheduling for application-level QoS delivery," *Proc. Protocols for Multimedia Systems - Multimedia Networking, 1997,* pp. 172 - 181, Santiago, Chile, Nov. 1997.

[AryrJo99] Y. Aryropoulos, S. Jordan, S.P.R. Kumar. "Dynamic Channel Allocation in Interference-Limited Cellular Systems with Uneven Traffic Distribution," *IEEE Transactions on Vehicular Technology,* vol. 48, iss. 1, pp. 224-232, Jan. 1999.

[AyyaCh05] D. Ayyagari, W. Chan, "Comparison of TDM and OFDMA access methods for powerline OFDM systems," *2$^{nd}$ IEEE Consumer Communications and Networking Conference, 2005. CCNC. 2005*, pp. 364- 368, Nevada, USA, Jan. 3-6, 2005.

[BahaSa99] A. R. S. Bahai, B. R. Saltzberg, *Multi-carrier digital communications, theory and applications of OFDM*, Kluwer Academic/Plenum, ISBN: ISBN 0-306-46296-6, New York, USA, 1999.

[Baum08] M. Baumann, *YATS-Simulator, User's and Programmer's Manual*, Dresden University of Technology, http://www.ifn.et.tu-dresden.de/tk/[2008].

[BordAn05] B. Bordbar, R. Anane, K. Okano, "An Evaluation Mechanism for QoS Management in Wireless Systems," *Proc. 11th Int. Conf. Parallel and Distributed Systems,* pp.150-154, Fuduoka, Japan, Jul. 20-22, 2005.

[Boud07] J-Y. Le Boudec, "Rate adaptation, congestion control and fairness: a tutorial," *http://ica1www.epfl.ch/PS_files/LEB3132.pdf*, Nov. 14, 2007.

[BoukHo00] A. Boukerche, S. Hong, T. Jacob. "Distributed Dynamic Channel Allocation for Mobile Communication Systems, Parallel Simulations and Distributed Systems Research," *Int. Symposium on Modelling*, pp. 73-81, 2000.

[BubMa07] M. Bublin, M. Konegger, P. Slanina. "A Cost-Function-Based Dynamic Channel Allocation and Its Limits," *IEEE Transactions on Vehicular Technology,* vol. 56, iss. 4, pp. 2286-2295, July 2007.

[Chan00] W. C. Chan, *Performance analysis of Telecommunications and local area networks*. Kluwer Academic Publishers Group, Netherlands, 2000.

[ChenSe01] J. Chen, D. Seah, W. Xu, "*Channel Allocation for Cellular Networks Using Heristic Methods*", Cornell University, May, 2001. <http://foulard.ee.cornell.edu/wxu/projects/cs574_project.html>

[D11OP07] Deliverable D11, "Reference guide on the description of field trials and test procedures." *The OPERA Consortium*, Jun. 2007.

[D13OP07] Deliverable D13, "Reference Guide on the Design of an Integrated PLC Network, Including the adaptations to allow the Carriers' Carrier model", *The OPERA Consortium*, Mar. 2007.

[D15OP07] Deliverable D15, "Results of the performed tests." *The OPERA Consortium*, Dec. 2008.

[D16OP05] Deliverable D16, "Reference guide for the selection of PLC distribution network technology", *The OPERA Consortium*, Mar. 2005.

[D18OP05] Deliverable D18, "Specification for a coexistence mechanism (access /in-home PLC)", *The OPERA Consortium*, Dec. 2005.

[D22OP08] Deliverable D22, "Report on the implementation MAC level OPERA PLC network simulator", *The OPERA2 Consortium*, Apr. 2008.

[D28OP07] Deliverable D28, "Improved coexistence specification", *The OPERA2 Consortium*, Mar. 2007.

[D51RE03] Deliverable 5.1, *Physical Layer Emulator,* The REMPLI consortium, 2003.

[DamoSi99] S. Damodaran, K.M. Sivalingam, "Scheduling in wireless networks with multiple transmission channels", *Proceedings of 7th Int. Conference on Network Protocols (ICNP '99)*, pp. 262 – 269, Toronto, Canada, Oct. 31-Nov. 3, 1999.

[DebPr02] K. Deb, A. Pratap, S. Agarwal, T. Meyarivan, "A fast and elitist multiobjective genetic algorithm: NSGA-II," *IEEE Trans. on Evolutionary Computation,* vol.6, iss.2, pp.182-197, Apr. 2002.

[DoHr05] L. P. Do, H. Hrasnica, and G. Bumiller, "Investigation of MAC Protocols for Single Frequency Network Technique Applied in Powerline Communications", *9th Int. Symposium on Power-Line Communications and Its Applications (ISPLC 2005)*, pp. 22 – 26, Vancouver, Canada, Apr. 2005.

[DoHr06] L. P. Do, H. Hrasnica, and G. Bumiller, "SALA MAC Protocol for PLC Networks based on Single Frequency Network Technique", *10th IEEE Int. Symposium on Power-Line Communications and Its Applications (IEEE-ISPLC 2006)*, pp. 172 – 177, Orlando, FL, USA, Mar. 26-29, 2006.

[DoHr07] L. P. Do, H. Hrasnica, and G. Bumiller, "CATP: Collective Acknowledgement Transmission Protocol for Group Commands in PLC Networks based on the SFN Technique", *11th IEEE Int. Symposium on Power-Line Communications and Its Applications (IEEE-ISPLC 2007)*, pp. 453-458, Pisa, Italy, Mar. 26-28, 2007.

[DoHr07a] L. P. Do, H. Hrasnica, and R. Lehnert, "Performance Evaluation of the PLC-MAC Protocol in accordance with the OPERA Specification", *11th IEEE Int. Symposium on Power-Line Communications and Its Applications (IEEE-ISPLC 2007)*, pp. 447-452, Pisa, Italy, Mar. 26-28, 2007.

[DoLe08] L. P. Do, R. Lehnert, "Scheduling Strategies for Service Admission in Powerline Communication Access Networks with QoS Support", *12th IEEE Int. Symposium on Power-Line Communications and Its Applications (IEEE-ISPLC 2008)*, pp. 41 – 46, Jeju, Korea, Apr. 2-4, 2008.

[DoLe09] L. P. Do, R. Lehnert, "Scheduling Strategies support for Multiple VPNs over Broadband PLC Access Network", *ITC Specialist Seminar 20*, Hoian, Vietnam. 18-20 May. 2009. <http://www.itcspecialistseminar.com/paper/itcss09_Do.pdf>

[Dost01] K. Dostert. *Powerline Communications*. Prentice Hall PTR, 2001.

[FeleHu07] M. Felegyhazi, J. P. Hubaux, Game Theory in Wireless Networks: A Tutorial. *EPFL–Switzerland.* 2007. <http://infoscience.epfl.ch/record/79715/files/>

[FezzRe06] K. Fezzani, C. Rebai, A. Ghazel, "Analysis and optimization of power line coupling circuit for CENELEC-PLC Modem," *13th IEEE Int. Conference on Electronics, Circuits and Systems, 2006. ICECS '06.* pp. 676-679, Dec. 10-13, 2006.

[FonsFl93] C.M. Fonseca and P.J. Fleming. "Genetic algorithms for multi objective optimization: Formulation, discussion and generalization," *Proc. of 5th Int. Conference on Genetic Algorithms*, pp. 416-423, San Mateo, USA, Jun. 1993.

[FranNi97] F. D. Priscoli, N. P. Magnani, V. Palestini, F. Sestini. "Application of Dynamic Channel Allocation Strategies to the GSM Cellular Network," *IEEE Journal on Selected Areas in Communications,* vol. 15, iss. 8, pp. 1558-1567. Oct.1997.

[GallKu09] S. Galli, A. Kurobe, "The Inter-PHY Protocol (IPP): A simple coexistence protocol for shared media," *Proc. of IEEE–ISPLC2009*, pp. 194 - 200, Dresden, Germany, Mar. 29-31, 2009.

[GambYu04] V. Gambiroza, P. Yuan, L. Balzano, Y. Liu, S. Sheafor, and E. Knightly, "Design, analysis, and implementation of DVSR: a fair high performance protocol for packet rings," *IEEE/ACM Trans. Networking*, vol. 12, iss. 1, pp. 85-102, Feb. 2004.

[GregSh00] M.H. MacGregor, W. Shi ,"Deficits for bursty latency-critical flows: DRR++", *Proc. of ICON2000*, pp. 287 – 293, Singapore, Sept. 5-8, 2000.

[GuLu07] X. Gu, A. Lukoyanov, J. Cao, "On the Evaluation of Traffic Scheduling and QoS for Wireless Multimedia Applications," *Inter. Conf. Wireless Communications, Networking and Mobile Computing, WiCom 2007,* pp.1962-1966, Shanghai, China, Sept. 21-25, 2007.

[Haid07] A. Haidine, R. Lehnert: "Analysis of the Channel Allocation Problem in Broadband Power Line Communications Access Networks" *Proc. ISPLC 2007*, pp. 192 – 197, Pisa, Italy, Mar. 26-28, 2007.

[Haid08] A. Haidine, R. Lehnert: "Solving the Generalized Base Station Placement problem in the planning of Broadband Power Line Communications access networks." *Proc. ISPLC 2008,* pp. 1-8, Jeju Island, South Korea, April 2-4, 2008.

[HoanJo02] H. Hoang, M. Jonsson, U. Hagstrom, A. Kallerdahl, "Switched real-time Ethernet with earliest deadline first scheduling protocols and traffic handling," *Int. Parallel and Distributed Processing Symposium, IPDPS 2002,* pp. 94-99, Florida, USA, Apr.15-19. 2002

[Holt01] J.M. Holtzman, "Asymptotic analysis of proportional fair algorithm," *12th IEEE - PIMRC2001*, pp. 33-37, California, USA, Sep. 30-Oct. 03, 2001.

[HomePa07] Merged of HomePlug and Panasonic as Draft Standard for Broadband over Power Line Networks: Medium Access Control and Physical Layer Specifications, Release 037, Sept. 2007. (unpublished)

[HondOh05] O. Honda, H. Ohsaki, M. Imase, J. Murayama, and K. Matsuda, "Scalable IP-VPN flow control mechanism supporting arbitrary fairness criteria — part 1: Architecture design," *in Proc. 14th Int. Conf. on Computer Communications and Networks (ICCCN 2005)*, pp. 173 – 178, San Diego, USA, 17-19 Oct. 2005.

[Hras04] H. Hrasnica, "Medium Access Control Protocols for Powerline Communications Network", Dissertation, Germany, May 2004.

[HrasHa04] H. Hrasnica, A. Haidine, R. Lehnert, *Broadband Powerline Communications Networks - Network Design*, John Wiley & Sons Inc.; June 2004.

[IshiMi84] H. Ishio, J. Minowa, K. Nosu, "Review and status of wavelength-division-multiplexing technology and its application," *IEEE Journal of Lightwave Technology*, vol.2, no.4, pp. 448- 463, Aug. 1984.

[IshiNo06] H. Ishibushi, Y. Nojima, and T. Doi. "Comparison between single-objective and multi-objective genetic algorithms: Performance comparison and performance measures." *Proc. of the 2006 IEEE Congress on Evolutionary Computation*, pp. 1143–1150, Vancouver, Canada, Jul. 16-21, 2006.

[ITU-G.711] "G.711: Pulse code modulation (PCM) of voice frequencies", *ITU-T Recommendation (11/1988),* http://www.itu.int/rec/T-REC-G.711/ 07.11.2011

[ITUT08] ITU-T-E.800, "Quality of services; concepts, models, objectives, dependability planning - Terms and definitions related to the quality of telecommunication services", revision 2008.

[JainCh84] R. Jain, D. Chiu, and W. Hawe, "A Quantitative Measure of Fairness and Discrimination for Resource Allocation in Shared Computer Systems." *DEC Research Report TR-301,* Digital Equipment Corporation, Maynard, MA, 1984.

[JonaGu02] H. J. Chao, X. Guo, *Quality of Service Control in High-Speed Networks*, John Wiley & Sons Inc, 2002.

[Kana92] T. Kanai, "Autonomous reuse partitioning in cellular systems," *in Proc. IEEE VTC*, pp. 782–785, Denver, USA, May 10-13,1992.

[KellMa98] F. P. Kelly, A. K. Maulloo and D. K. H. Tan., "Rate control in communication networks: shadow prices, proportional fairness and stability," *Journal of the Operational Research Society*, vol. 49, pp. 237-252, Apr. 1998.

[KirkGe83] S. Kirkpatrick, C.D. Gelatt. "Optimization by Simulated Annealing", *Science*, vol. 220, No. 4598, pp. 671-680, May 1983.

[KwonSe09] H. Kwon; H. Seo; S. Kim; B. G. Lee, "Generalized CSMA/CA for OFDMA systems: protocol design, throughput analysis, and implementation issues," *IEEE Transactions on Wireless Communications*, vol.8, no.8, pp.4176-4187, Aug. 2009.

[Laur96] A. Lauralee, "Quality of experience: defining the criteria for effective interaction design", *Magazine Interactions*, vol. 3, iss. 3, pp. 11-15, May/June 1996.

[LeeUn86] H. Lee, C. Un "*A Study of On-Off Characteristics of Conversational Speech*", IEEE Trans. Communications, vol. 34, iss. 6, pp. 630–637, 1986.

[Lehn09] R. Lehnert, *Modeling and Simulation of Telecommunication Systems and Networks*, Lecture Script for WS08/09, Technische Universität Dresden, revision 25.01.2009.

[LeiLi09] L. Lei, C. Lin, J. Cai, X. Shen, "Performance Analysis of Wireless Opportunistic Schedulers using Stochastic Petri Nets," *IEEE Trans. on Wireless Communications*, vol. 8, no.4, pp. 2076-2087. Apr. 2009

[LiniDo05] N. Lilith, K. Dogancay, "Distributed Dynamic Call Admission Control and Channel Allocation Using SARSA" *Asia Pacific Conference on Communications,* pp. 376-380, Perth, WA, Oct. 3-5, 2005.

[LoiWi99] B. Lóic, R. Wilfrid, "Dynamic Channel Assignment in GSM networks," *50th Vehicular Technology Conference*, pp. 2339-2342. Amsterdam, Netherlands, Sept. 19-22, 1999.

[LozaDo02] A. Lozano, C. C. Donald, "Distributed Dynamic Channel Assignment in TDMA Mobile Communication Systems," *IEEE Transactions on Vehicular Technology*, vol. 51, iss. 6, pp. 1397-1406, Nov. 2002.

[LuoLi01] W. Luo; C. Lin; B. Yan, "A congestion control mechanism supporting QoS requirements," *Proc. IEEE Int. Conf. Computer Networks and Mobile Computing*, pp. 55-60, Beijing, China, Oct. 16-19, 2001.

[MastPa05] R. Mastrodonato, G. Paltenghi, "Analysis of a Bandwidth Allocation Protocol for Ethernet Passive Optical Networks (EPONs)", *ICTON 2005*, . pp. 241-244, Barcelona, Spain, July 3-7, 2005.

[MeshCh05] F. Meshkati, M. Chiang, S. C. Schwartz, H. V. Poor. "A Non-Cooperative Power Control Game for Multi-Carrier CDMA Systems," *IEEE-Wireless Comm. Networking Conf.,* pp. 606-611. LA, USA, Mar. 13-17, 2005

[MoufNa06] H. T. Mouftah, H. Naser, "A joint-ONU interval-based dynamic scheduling algorithm for Ethernet Passive Optical Networks," *IEEE/ACM transactions on networking*, vol. 14, no. 4, pp. 889–899, Aug. 2006.

[NiyaHo07] D. Niyato, E. Hossain, "Optimal Price Competition for Spectrum Sharing in Cognitive Radio: A Dynamic Game-Theoretic Approach," *Globecom 2007,* pp. 4625 – 4629, Washington DC, USA, 2007.

[Opera06] "OPERA Specification Part 1: Technology, Part 2: System", *The OPERA Consortium*, Jan. 2006.

[Ozda08] A. Ozdaglar, "Price Competition with Elastic Traffic", *Journal Networks - Games, Interdiction, and Human Interaction Problems on Networks,* vol.52, iss.3, pp. 141-155, Oct. 2008

[PaceAl05] P. Pace, G. Aloi, S. Marano, "A multilayered architecture supporting QoS for multimedia traffic connections", *IEEE 61st Vehicular Technology Conference,* vol. 4, pp 2702 - 2706. Stockholm, Sweden, May 30-Jun. 01, 2005.

[PaoLa98] D.C.W. Pao, S.P. Lam, "Cell scheduling for ATM switch with two priority classes", *IEEE-ATM Workshop Proceedings*, pp. 86-90, VA, USA, May 26-29, 1998.

[PapzKa05] P. M. Papzoglou, D. A. Karras, R. C. Papademetriou, "A dynamic channel assignment simulation system for large scale cellular telecommunications", *Int. Conference HERCMA 2005,* pp. 45-52, Athens, Greece, Sept. 2005.

[QinLi02] Y. Qin, B. Li, W. Lim, S.H. Yeo, "A design for on-line virtual private networks (VPN) over optical WDM networks", *IEEE-Globecom2002*, pp. 2782 – 2786, Taipei, Nov. 17-21, 2002.

[RaduBo07] B. Radunovic, J-Y. Le Boudec, "A Unified Framework for Max-Min and Min-Max Fairness With Applications," *IEEE/ACM Transactions on Networking,* vol.15, no.5, pp.1073-1083, Oct. 2007.

[SalmGi01] M. Salmenkaita, J. Gimenez, P. Tapia. "A Practical DCA Implementation for GSM Networks: Dynamic Frequency and Channel Assignment," *Vehicular Tech. Conf.,* pp. 2529-2533, Rhodes, Greece, May 6-9, 2001,

[Sche02] J. Scheffler, "*Bestimmung der maximal zulässigen Netzanschlussleistung photovoltaischer Energiewandlungsanlagen in Wohnsiedlungsgebieten*", Dissertation for the Degree of Doctor, TU Chemnitz, 2002. (In German)

[ShinCh99] S. M. Shin, C. H. Cho, and D. K. Sung, "Interference -based channel assignment for DS-CDMA cellular systems," *IEEE Trans. Veh. Technology*, vol. 48, no. 1, pp. 233–239, Jan. 1999.

[ShreVa96] M. Shreedhar, G. Varghese, "Efficient Fair Queuing Using Deficit Round-Robin", IEEE/ACM Trans. Net., vol. 4, no. 4, pp. 386-397. Jun. 1996.

[SorePo05] T. B. Sorensen, M. R. Pons, "Performance Evaluation of Proportional Fair Scheduling Algorithm with Measured Channels", *IEEE 62nd Vehicular Technology Conf.*, pp. 2580 - 2585, Dallas, USA, Sept. 2005.

[Stin02] Maxwell B. Stinchcombe. *Notes for a Course in Game Theory*. The MIT Press, Massachusetts Institute of Technology Cambridge, 2002.

[SumaKu06] B. Suman, P. Kumar. "A survey of simulated annealing as a tool for single and multiobjective optimization." *Journal of Operational Research Society*. pp. 1143-1160. 2006.

[SuriDa07] J. E. Suris, L. A. DaSilva, Z. Han, A. B. MacKenzie. "Cooperative Game Theory for Distributed Spectrum Sharing," *IEEE Int. Conference on Communications-ICC2007*, pp. 5282 – 5287, Glasgow, Scotland, Jun. 24-28, 2007.

[ToneZh09] A. M. Tonello, T. Zheng, "Bottom-up transfer function generator for broadband PLC statistical channel modeling," *Proc. of IEEE Int. Sym. Power Line Communications and Its Applications(ISPLC2009)*, pp.7-12, Dresden, Germany, Mar. 29-April 1, 2009.

[Torr08] B. M. Torres, "*Investigation of Real-Time Scheduling for Broadband PLC Networks*", master thesis, 2008.

[TsaiCh10] T.Y. Tsai, Y. L. Chung, Z. Tsai, *Introduction to Packet Scheduling Algorithms for Communications Networks*, published by InTech - Open Access Publisher, ISBN 978-953-307-114-5, Sept. 2010.

[TseHa98] D. N. C. Tse, S. V. Hanly. "Multiaccess Fading Channels–Part I: Polymatroid Structure, Optimal Resource Allocation and Throughput Capacities", *IEEE Trans. on Information Theory*, vol. 44, no. 7, pp. 2796-2815, Nov. 1998.

[VandFa00] B. Vandalore, S. Fahmy, R. Jain, R. Goyal, and M. Goyal, "General weighted fairness and its support in explicit rate switch algorithms,"

*Computer Communication Mag.*, vol. 23, pp. 149-161, Jan. 2000.

[VogeKe95] A. Vogel, B. Kerherve, G. Bochmann, J. Gecsei, "Distributed multimedia and QOS: a survey," *IEEE Multimedia,* vol.2, no.2, pp.10-19,Summer 1995.

[PiamVi09] K. Piamrat, C.Viho, J.-M. Bonnin, A. Ksentini, "Quality of Experience Measurements for Video Streaming over Wireless Networks," $6^{th}$ *Int. Conf. on Info. Tech: ITNG '09.* pp.1184-1189, Nevada, USA, April 27-29, 2009

[Welz05] M. Welzl, *Network Congestion Control - Managing Internet Traffic*, John Wiley & Sons Inc.; ISBN-13 978-0-470-02528-4, 2005.

[WenCh01] J. Wen, W. J. Chen, S. Y. Lin, K. T. Huang, "Performace evaluation of LIBTA/hybrid time-slot selection algorithm for cellular systems," *Internal journal of communication systems*, vol.14, pp. 575-591, 2001.

[WidmDe01] J. Widmer, R. Denda,M. Mauve, "A survey on TCP-friendly congestion control," *IEEE Network* , vol.15, no.3, pp.28-37, May 2001.

[WiMAX04] IEEE 802.16-2004, "IEEE Standard for Local and Metropolitan Area Networks - Part 16: Air Interface for Fixed Broadband Wireless Access Systems," Oct. 2004.

[Wong03] S. H. Wong, "*Channel allocation for Broadband fixed wireless access networks*", Dissertation for the Degree of Doctor, University of Cambridge, 2003.

[WongWa02] S. H. Wong, I.J. Wassell, "Application of game theory for distributed dynamic channel allocation," *IEEE* $55^{th}$ *Vehicular Technology Conference,* pp. 404 – 408, Alabama, USA, May 06-09, 2002

[ZhanHe05] Z. Zhang, Y. He, E. K. P. Chong, "Opportunistic downlink scheduling for multiuser OFDM systems". *WCNC2005*: *IEEE-Wireless Communications and Networking Conf.* pp. 1206 – 1212, LA, USA, Mar. 2005.

[ZhonXu05] W. Zhong, Y. Xu, Y. Cait, "Capacity and Game-Theoretic Power Allocation for Multiuser MIMO Channels with Channel Estimation Error," *IEEE Int. Symp. on Communications and Information Tech.* , pp. 716-719, Beijing, China, Oct. 12-14, 2005.

# A Appendix

## A. 1 Procedures and Parameters from OPERA Specification

The parameters presented in this appendix are extracted from the [Opera06] specification.

### A.1.1 Data Frame Structure

Each data frame consist of one or several bursts, each burst addresses to a users. Inside a burst, several OPERA packets are transmitted. The OPERA packet is equivalent to the Ethernet packet with a slightly modified packet header.

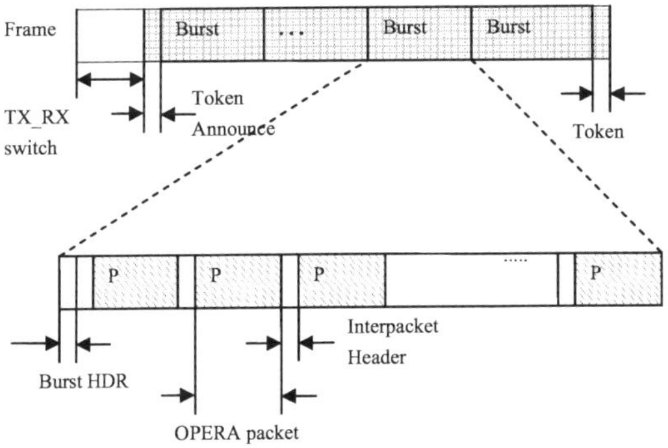


HDR: header
P: OPERA packet
TX/RX switch: interval between transmission and reception

*Figure A-1: OPERA data frame structure*

### A.1.2 Transmission Sequences: BS-Users (without repeater)

***Mode 1: without time frame structure***

The BS transmits data to and receives data from each user in a sequence. It is suggested to be used with low load network situations. At the end of each transmission, a data token is sent to the node which is allowed to transmit in the next period.

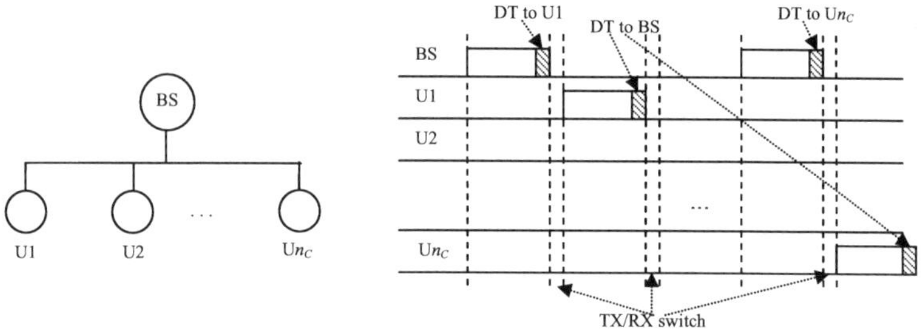


DT: data token
TX/RX switch: interval between transmission and reception

*Figure A-2: Transmission Sequences: BS-Users in Mode 1*

***Mode 2: with time frame structure***

Each time frame (TF) is subdivided into eight sub time frame (sTF). In the first part of a sub time frame, after transmitting to all users, the BS transmits a distributor token to all users. This distributor token consists of the time allocation for uplink transmissions (from the users to the BS).

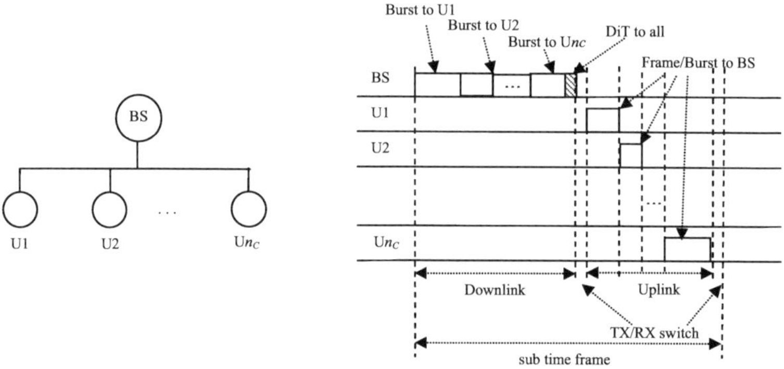


DiT: Distributor token
TX/RX switch: interval between downlink and uplink transmissions

*Figure A-3: Transmission Sequences: BS-Users in Mode 2*

### A.1.3 Transmission Sequences: BS-TDRs-Users

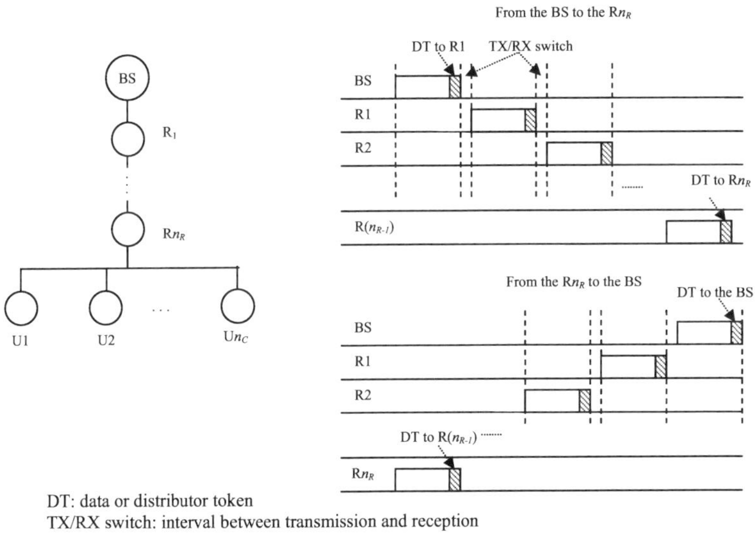


*Figure A-4: Transmission Sequences: BS- TDRs-Users*

### A.1.4 OPERA Parameters

| | | |
|---|---|---|
| $T_{Sym}$ | 71µs | OFDM Symbol duration |
| $S^{PHY}$ | 200Mbps | Data rate in physical layer |
| $N_{sTF}$ | 422 symbols | Number of symbols in a sub time frame (sTF) |
| $N_{AN}$ | 1 symbol | Duration of a token announcement |
| $N_{DT}$ | 1 symbol | Duration of a data/distributor token |
| $N_{OV}^{1B}$ | 1 symbol | Overhead of a burst |

*Table A-1: OPERA System parameters*

| | | |
|---|---|---|
| $L^{PL}$ | 42-1500 bytes | Length of a OPERA packet payload |
| $L^{POV}$ | 34 bytes | Overhead of a OPERA packet |
| $L^{IH}$ | 0 or 4 bytes | Inter header between OPERA packets |

*Table A-2: OPERA Packet parameters*

### A.1.5 Additional Procedures

In order to organize the transmissions in the PLC network, several protocols and procedures have been implemented. Three important protocols and procedures are reviewed in this section.

#### A.1.5.1 Synchronization

The so-called "announce token" is generated and broadcasted regularly from the BS at the beginning of each time frame. This message is also repeated by the repeaters to the users. It contains information about the network, e.g. network ID, configuration parameters, as well as the necessary information for synchronization. When a node receives this message, it synchronizes its clock with the received one from its master.

#### A.1.5.2 Automatic Master Selection Protocol

When a node joins the network, it has to listen for the announce token from the BS in the PLC network. If several tokens were received, the node calculates the transmission data rates to the BSs based on the received channel information. After that, the new node registers directly with one of the BSs or via a repeater to form a logical connection which results in the highest transmission data rate.

#### A.1.5.3 Channel Measurement Procedure

In order to determine or update the data rate of a connection, the specification in [Opera06] defines a special procedure called channel measurement procedure. This procedure uses a special "channel measurement frame" sent from the transmitter to the receiver to measure the connection. When this frame is received, the receiver estimates the number of bits per carrier (BPC) in each carrier and creates the so-called bit-loading table to send back to the transmitter. With the received bit-loading table from the receiver, the transmitter updates the transmission capacity for this connection before sending the data. The channel measurement scheme is presented in Fig. A-5.

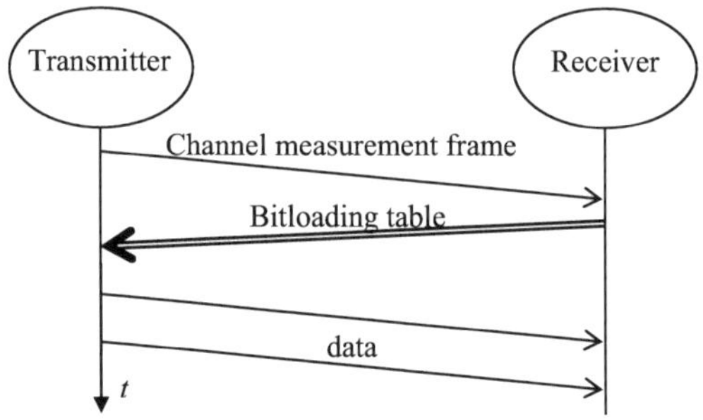


*Figure A-5: PLC channel measurement scheme*

## A. 2 Proofs of Mathematic Expressions

### A.2.1 Limit values of the average delivery time and average throughput of the network in chain structure

For a chain network structure of the PLC cells, when number of cells increase, the average delivery time of the CA-Msg is

$$\lim_{S\to\infty}(\overline{D_{Ch}}) = 6.75 \text{ and} \tag{8-1}$$

The average throughput of the CA-Msg in each cell is

$$\lim_{S\to\infty}(\overline{p^{recv}}) = {}^{8}\!/\!_{27} = 0.\overline{296} \tag{8-2}$$

Proof:

The objective is to calculate the limit values with large *S*. The following calculations applied for the network with *S*≥6. Considering the network in the chain structure with *S* cells as presented in Fig. A-6, the number of the neighbor cells of each cell ($H_i$) and probability of sending a CA-Msg in a time frame ( $p_i^{send}$ ) calculated by (6-31) are presented.

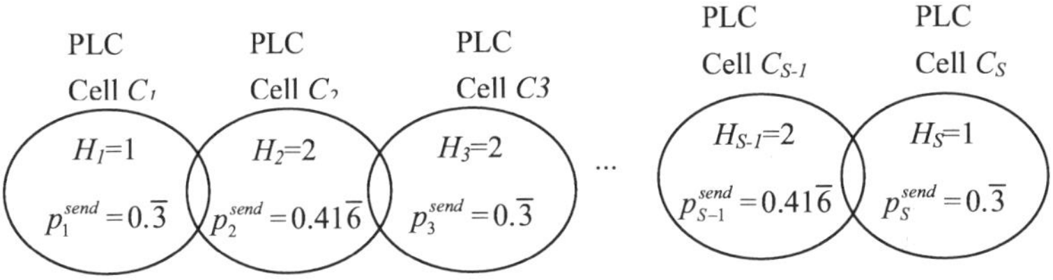


*Figure A-6: Number of neighbor and CA-Msg sending probability in PLC network in a chain structure*

In accordance with (6-31):

$$p_1^{send} = 1/(H_2+1) = 1/3 = 0.(3) = p_S^{send}$$

$$p_2^{send} = (1/H_2)(1/(H_1+1)+1/(H_3+1)) = 5/12 = 0.41\overline{6} = p_{S-1}^{send}$$

$$p_3^{send} = (1/H_3)(1/(H_2+1)+1/(H_4+1)) = 1/3 = 0.\overline{3} = p_4^{send} = \ldots = p_{S-2}^{send}$$

Denote $p_{j,i}^{recv}$ as $p_{j\to i}^{recv}$. In accordance with (6-25), probability of receiving CA-Msg from the neighbor:

$$p_{1,2}^{recv} = p_2^{send}(1-p_1^{send}) = 5/18 = 0.2\overline{7} = p_{S,S-1}^{recv}$$

$$p_{2,1}^{recv} = p_1^{send}(1-p_2^{send})(1-p_3^{send}) = 7/54 = 0.1\overline{296} = p_{S-1,S}^{recv}$$

$$p_{2,3}^{recv} = p_3^{send}(1-p_2^{send})(1-p_1^{send}) = 7/54 = 0.1\overline{296} = p_{S-1,S}^{recv}$$

$$p_1^{recv} = p_{1,2}^{recv} = p_S^{recv}$$

...

$$p_1^{recv} = p_{1,2}^{recv} = p_S^{recv}$$

$$p_2^{recv} = p_{2,1}^{recv} + p_{2,3}^{recv} = p_{S-1}^{recv}$$

$$p_3^{recv} = p_{3,2}^{recv} + p_{3,4}^{recv} = p_{S-2}^{recv}$$

From cell $C_4$ to cell $C_{S-3}$:

$$p_{4,3}^{recv} = p_3^{send}(1-p_4^{send})(1-p_5^{send}) = 4/27 = 0.\overline{148} = p_{5,4}^{recv} = ... = p_{S-2,S-3}^{recv}$$

$$p_{4,5}^{recv} = p_5^{send}(1-p_4^{send})(1-p_3^{send}) = 4/27 = 0.\overline{148} = p_{5,6}^{recv} = ... = p_{S-3,S-2}^{recv}$$

So that:

$$p_4^{recv} = p_{4,3}^{recv} + p_{4,5}^{recv} = 8/27 = p_5^{recv} = ... = p_{S-3}^{recv}$$

The network is in the chain structure, so the number of the connections between the cell is $n^{conn} = S-1$. In accordance with (6-27):

$$\overline{p^{recv}} = \frac{1}{S}\sum_{i=1}^{S} p_i^{recv} = \frac{p_1^{recv} + p_2^{recv} + p_3^{recv} + p_4^{recv} + ... + p_{S-3}^{recv} + p_{S-2}^{recv} + p_{S-1}^{recv} + p_S^{recv}}{S}$$

$$= \frac{p_1^{recv} + p_2^{recv} + p_3^{recv} + p_{S-2}^{recv} + p_{S-1}^{recv} + p_S^{recv}}{S} + \frac{p_4^{recv} + ... + p_{S-3}^{recv}}{S}$$

$$= \frac{p_1^{recv} + p_2^{recv} + p_3^{recv} + p_{S-2}^{recv} + p_{S-1}^{recv} + p_S^{recv}}{S} + \frac{(S-6)p_4^{recv}}{S}$$

Thus:

$$\lim_{S\to\infty}(\overline{p^{recv}}) = \lim_{S\to\infty}\left(\frac{p_1^{recv} + p_2^{recv} + p_3^{recv} + p_{S-2}^{recv} + p_{S-1}^{recv} + p_S^{recv}}{S}\right) + \lim_{S\to\infty}\left(\frac{(S-6)p_4^{recv}}{S}\right)$$
$$= p_4^{recv} = 0.\overline{296}$$

$$\overline{D_{Ch}} = \frac{1}{2n^{conn}} \sum_{i=1}^{S} \sum_{j=1}^{H_i} 1/p_{i,j}^{recv} = \frac{1}{2n^{conn}} \sum_{i=1}^{S} \left( \frac{1}{p_{1,2}^{recv}} + \frac{1}{p_{2,1}^{recv}} + \frac{1}{p_{2,3}^{recv}} + \frac{1}{p_{3,2}^{recv}} + \frac{1}{p_{3,4}^{recv}} + \ldots + \frac{1}{p_{S,S-1}^{recv}} \right)$$

$$= \frac{1}{2(S-1)} \left( \frac{1}{p_{1,2}^{recv}} + \frac{1}{p_{2,1}^{recv}} + \frac{1}{p_{2,3}^{recv}} + \frac{1}{p_{3,2}^{recv}} + \frac{1}{p_{3,4}^{recv}} \right)$$

$$+ \frac{1}{2(S-1)} \left( \frac{1}{p_{4,3}^{recv}} + \frac{1}{p_{4,5}^{recv}} + \ldots + \frac{1}{p_{S-3,S-4}^{recv}} + \frac{1}{p_{S-3,S-2}^{recv}} \right)$$

$$+ \frac{1}{2(S-1)} \left( \frac{1}{p_{S-2,S-3}^{recv}} + \frac{1}{p_{S-2,S-1}^{recv}} + \frac{1}{p_{S-1,S-2}^{recv}} + \frac{1}{p_{S-1,S}^{recv}} + \frac{1}{p_{S,S-1}^{recv}} \right)$$

$$= \frac{1}{2(S-1)} \left( \frac{1}{p_{1,2}^{recv}} + \frac{1}{p_{2,1}^{recv}} + \frac{1}{p_{2,3}^{recv}} + \frac{1}{p_{3,2}^{recv}} + \frac{1}{p_{3,4}^{recv}} \right)$$

$$+ \frac{1}{2(S-1)} \left( \frac{1}{p_{S-2,S-3}^{recv}} + \frac{1}{p_{S-2,S-1}^{recv}} + \frac{1}{p_{S-1,S-2}^{recv}} + \frac{1}{p_{S-1,S}^{recv}} + \frac{1}{p_{S,S-1}^{recv}} \right)$$

$$+ \frac{1}{2(S-1)} \left( \frac{2(S-3)}{p_{4,3}^{recv}} \right)$$

With S→∞:

$$\lim_{S\to\infty}(\overline{D_{Ch}}) = \lim_{S\to\infty} \frac{1}{2(S-1)} \left( \frac{1}{p_{1,2}^{recv}} + \frac{1}{p_{2,1}^{recv}} + \frac{1}{p_{2,3}^{recv}} + \frac{1}{p_{3,2}^{recv}} + \frac{1}{p_{3,4}^{recv}} \right)$$

$$+ \lim_{S\to\infty} \frac{1}{2(S-1)} \left( \frac{1}{p_{S-2,S-3}^{recv}} + \frac{1}{p_{S-2,S-1}^{recv}} + \frac{1}{p_{S-1,S-2}^{recv}} + \frac{1}{p_{S-1,S}^{recv}} + \frac{1}{p_{S,S-1}^{recv}} \right)$$

$$+ \lim_{S\to\infty} \frac{1}{2(S-1)} \left( \frac{2(S-3)}{p_{4,3}^{recv}} \right)$$

$$= \frac{1}{p_{4,3}^{recv}} = \frac{27}{4} = 6.75$$

### A.2.2 Limit values of the average delivery time and average throughput of the network in star structure

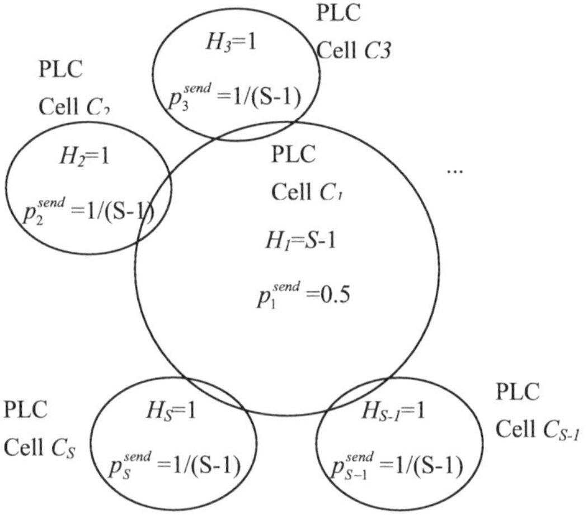


*Figure A-7: Number of neighbor and CA-Msg sending probability in PLC network in a star structure*

For a star network structure of the PLC cells, when number of cells increase, the average delivery time of the CA-Msg is

$$\lim_{S\to\infty}(\overline{D_{Ch}}) = \infty \text{ and} \tag{8-3}$$

The average throughput of the CA-Msg in each cell is

$$\lim_{S\to\infty}(\overline{p^{recv}}) = 0.5 \tag{8-4}$$

Proof:

Cell $C_1$ has $S$-1 neighbors ($H_1$=$S$-1). Each cell $C_2$ to $C_S$ has only one neighbor ($H_2$=$H3$=...=1). In accordance with (6-31), $C_1$ send its CA-Msg with the probability of:

$$p_1^{send} = \frac{1}{H_1}\cdot\left(\frac{1}{H_2+1}+..+\frac{1}{H_S+1}\right) = \frac{1}{S-1}\cdot\underbrace{\left(\frac{1}{2}+..+\frac{1}{2}\right)}_{S-1} = \frac{1}{2}$$

And from the cell $C_2$ to $C_S$

$$p_2^{send} = p_3^{send} =..= p_S^{send} = 1/(H_1+1) = 1/S$$

The reception of the CA-Msg for the cell $C_2$ .. $C_S$

$$p_2^{recv} = p_3^{recv} = ..p_S^{recv} = p_1^{send}(1 - p_2^{send}) = 0.5 \times \left(1 - \frac{1}{S}\right) = \frac{S-1}{2S}$$

And $C_1$:

$$p_1^{recv} = \sum_{i=2}^{S} p_{1,i}^{recv} = \sum_{i=2}^{S}\left[p_2^{send}(1 - p_1^{send})\right] = \sum_{i=2}^{S}\left[0.5 \times \frac{1}{S}\right] = \frac{S-1}{2S}$$

The delivery time for $C_1$ to $C_2$ .. $C_S$

$$D_{2,1} = D_{3,1} = .. = D_{S,1} = \frac{1}{p_2^{recv}} = \frac{2S}{S-1}$$

and $C_2$ .. $C_S$ to $C_1$:

$$D_{1,2} = D_{1,3} = .. = D_{1,S} = \frac{1}{p_{1,i}^{recv}} = \frac{1}{0.5/S} = 2S$$

When $S \to \infty$:

$$\lim_{S\to\infty} D_{2,1} = \lim_{S\to\infty} D_{3,1} = .. = \lim_{S\to\infty} D_{S,1} = \lim_{S\to\infty} \frac{2S}{S-1} = 2$$

$$\lim_{S\to\infty} D_{1,2} = \lim_{S\to\infty} D_{1,3} = .. = \lim_{S\to\infty} D_{1,S} = \lim_{S\to\infty} 2S = \infty$$

Therefore:

$$\lim_{S\to\infty}(\overline{D_{Ch}}) = \infty \text{ and}$$

$$\lim_{S\to\infty} p_1^{recv} = \lim_{S\to\infty} p_2^{recv} = \lim_{S\to\infty} p_3^{recv} = .. = \lim_{S\to\infty} p_S^{recv} = \lim_{S\to\infty} \frac{S-1}{2S} = 0.5$$

Therefore:

$$\lim_{S\to\infty}(\overline{p^{recv}}) = 0.5$$

## A. 3 Selected Simulation Results

A simulation result is derived from 10 simulation sub-runs, each lasting for 100.000 time frames. The statistical accuracy is set to the confidence intervals with a confidence probability of 0.95. The different of value *y* to value *x* (in %) is calculated by: $d=(y-x)/x\times100\%$.

PLC Cell $C_1$

| | **RRS** (Mean) | Confidence Interval | Different with RSS / % |
|---|---|---|---|
| Fitness *G* / BU | 161.122 | [156.718, 165.526] | |
| $B_U$ /BU | 85.0629 | [82.811, 87.315] | |
| $F_U$ | 0.8931 | [0.884, 0.903] | |
| Execution time /µs | 1.4292 | [1.404, 1.454] | |
| | **BBS** (Mean) | | |
| Fitness *G* / BU | 188.035 | [183.295, 192.775] | 16.70 |
| $B_U$ /BU | 99.3207 | [96.769, 101.872] | 16.76 |
| $F_U$ | 0.8925 | [0.887, 0.898] | -0.07 |
| Execution time /µs | 42.222 | [41.497, 42.948] | 2854.21 |
| | **HA** (Mean) | | |
| Fitness *G* / BU | 194.641 | [189.539, 199.742] | 20.80 |
| $B_U$ /BU | 95.415 | [92.974, 97.857] | 12.17 |
| $F_U$ | 0.9372 | [0.934, 0.941] | 4.94 |
| Execution time /µs | 846.453 | [814.573, 878.332] | 59124.41 |

*Table A-3: Comparisons of RRS, BBS, and HA algorithms with TDMA, $C_1$*

PLC Cell $C_2$

| | **RRS** (Mean) | Confidence Interval | Different with RSS / % |
|---|---|---|---|
| Fitness *G* / BU | 503.618 | [494.667, 512.568] | |
| $B_U$ /BU | 267.207 | [262.619, 271.796] | |
| $F_U$ | 0.884 | [0.878, 0.890] | |
| Execution time /µs | 3.331 | [3.312, 3.352] | |
| | **BBS** (Mean) | | |
| Fitness *G* / BU | 627.151 | [616.816, 637.486] | 24.53 |
| $B_U$ /BU | 329.575 | [324.459, 334.691] | 23.34 |
| $F_U$ | 0.9022 | [0.898, 0.906] | 2.06 |

| | | | |
|---|---|---|---|
| Execution time /ms | 0.431 | [0.424, 0.438] | 12858.74 |
| | **HA** (Mean) | | |
| Fitness *G* / BU | 631.567 | [621.611, 641.522] | 25.41 |
| $B_U$ /BU | 323.028 | [318.223, 327.834] | 20.89 |
| $F_U$ | 0.9230 | [0.920, 0.926] | 4.42 |
| Execution time /ms | 8.102 | [7.974, 8.231] | 243110.72 |

*Table A-4: Comparisons of RRS, BBS, and HA algorithms with TDMA, $C_2$*

PLC Cell $C_3$

| | **RRS** (Mean) | Confidence Interval | Different with RSS / % |
|---|---|---|---|
| Fitness *G* / BU | 2402.29 | [2385.510, 2419.070] | |
| $B_U$ /BU | 1277.6 | [1269.010, 1286.200] | |
| $F_U$ | 0.8801 | [0.879, 0.881] | |
| Execution time /μs | 14.1937 | [14.162, 14.225] | |
| | **BBS** (Mean) | | |
| Fitness *G* / BU | 2791.84 | [2775.180, 2808.510] | 16.22 |
| $B_U$ /BU | 1480.84 | [1472.620, 1489.070] | 15.91 |
| $F_U$ | 0.8852 | [0.884, 0.887] | 0.58 |
| Execution time /ms | 10.55 | [10.495, 10.611] | 74254.82 |
| | **HA** (Mean) | | |
| Fitness *G* / BU | 2787.98 | [2771.810, 2804.150] | 16.06 |
| $B_U$ /BU | 1420.22 | [1412.420, 1428.020] | 11.16 |
| $F_U$ | 0.9196 | [0.919, 0.921] | 4.48 |
| Execution time /ms | 205.686 | [204.816, 206.557] | 1449035.88 |

*Table A-5: Comparisons of RRS, BBS, and HA algorithms with TDMA, $C_3$*

| | HA | | HAAS | | Different /% |
|---|---|---|---|---|---|
| | Mean | Confidence Interval | Mean | Confidence interval | |
| *10 iterations* | | | | | |
| Fitness *G* / BU | 2523.46 | [2507.890, 2539.020] | 2828.28 | [2811.240, 2845.320] | 12.07 |
| $B_U$ /BU | 1323.77 | [1315.190, 1332.360] | 1469.58 | [1460.640, 1478.530] | 11.01 |

| | | | | | |
|---|---|---|---|---|---|
| $F_U$ | 0.9061 | [0.904, 0.908] | 0.9207 | [0.920, 0.921] | 1.60 |
| Execution time /ms | 11.7125 | [11.588, 11.837] | 12.4699 | [12.298, 12.642] | 6.46 |
| *100 iterations* | | | | | |
| Fitness $G$ / BU | 2787.98 | [2771.810, 2804.150] | 2849.34 | [2838.710, 2859.970] | 2.20 |
| $B_U$ /BU | 1420.22 | [1412.420, 1428.020] | 1478.29 | [1471.980, 1484.610] | 4.08 |
| $F_U$ | 0.9196 | [0.919, 0.921] | 0.9238 | [0.924, 0.924] | 0.46 |
| Execution time /ms | 205.686 | [204.816, 206.557] | 118.871 | [116.244, 121.499] | -42.20 |

*Table A-6: HA and HAAS with TDMA comparisons*

| | BBS-TDMA | | EBBS-TDMA/OFDMA | | Different /% |
|---|---|---|---|---|---|
| | Mean | Confidence Interval | Mean | Confidence interval | |
| *PLC cell $C_1$* | | | | | |
| Fitness $G$ / BU | 188.035 | [183.295, 192.775] | 246.179 | [237.873, 254.485] | 30.92 |
| $B_U$ /BU | 99.320 | [96.769, 101.872] | 124.885 | [120.81, 128.96] | 25.74 |
| $F_U$ | 0.892 | [0.887, 0.897] | 0.970 | [0.966, 0.973] | 8.73 |
| Execution time /ms | 0.042 | [0.041, 0.043] | 0.365 | [0.345, 0.384] | 764.69 |
| *PLC cell $C_2$* | | | | | |
| Fitness $G$ / BU | 627.151 | [616.816, 637.486] | 817.525 | [810.007, 825.043] | 30.36 |
| $B_U$ /BU | 329.575 | [324.459, 334.691] | 412.419 | [408.697, 416.142] | 25.14 |
| $F_U$ | 0.9022 | [0.899, 0.906] | 0.9819 | [0.980, 0.983] | 8.84 |
| Execution time /ms | 0.431 | [0.424, 0.438] | 11.146 | [10.992, 11.300] | 2481.83 |
| *PLC cell $C_3$* | | | | | |
| Fitness $G$ / BU | 2791.84 | [2775.18, 2808.51] | 3486.69 | [3363.05, 3610.32] | 24.89 |
| $B_U$ /BU | 1480.84 | [1472.62, 1489.07] | 1751.68 | [1687.48, 1815.88] | 18.29 |
| $F_U$ | 0.8852 | [0.883, 0.886] | 0.9907 | [0.987, 0.994] | 11.91 |
| Execution time /ms | 10.553 | [10.495, 10.611] | 814.934 | [810.475, 819.392] | 7621.78 |

*Table A-7: EBBS with TDMA/OFDMA and BBS with TDMA comparisons*

## A. 4 List of Abbreviations

| | |
|---|---|
| AMR | Automatic Meter Reading |
| ARQ | Automatic Repeat reQuest |
| BB-PLC | Broadband Powerline Communications |
| BBS | Benefit Based Selection Algorithm |
| BER | Bit Error Rate |
| BN | Border Node |
| BPC | Bit Per Carrier |
| BPSK | Binary Phase Shift Keying |
| BS | Base Station |
| BU | Benefit Unit |
| CAC | Connection Admission Control |
| CA-Msg | Channel Announcement Message |
| CA-TS | Channel Announcement Time Slot |
| CB | Cable Box |
| CDMA | Code Division Multiple Access |
| CENELEC | European Committee for Electrotechnical Standardization |
| CHT*i* | Channel quality model type *i* |
| CI | Confidence Interval |
| CSK | Code Shift Keying |
| CSMA | Carrier Sense Multiple Access |
| CSMA/CA | Carrier Sense Multiple Access with Collision Avoidance |
| CSMA/CD | Carrier Sense Multiple Access with Collision Detection |
| DCA | Dynamic Channel Allocation |
| DICRA | Distributed Inter-Cell Resource Allocation |
| DSL | Digital Subscriber Line |
| EMC | Electro Magnetic Compatibility |
| ES | Exhaustive Search |

| | |
|---|---|
| FCA | Fixed Channel Allocation |
| FCC | Federal Communications Commission |
| FCFS | First Come First Serve |
| FDM | Frequency Division Multiplexing |
| FDMA | Frequency Division Multiple Access |
| FDR | Frequency Division Repeater |
| FEC | Forward Error Correction |
| FFT | Fast Fourier Transform |
| FSK | Frequency Shift Keying |
| GA | Genetic Algorithm |
| GSM | Global System for Mobile communication |
| HA | Heuristic Algorithm |
| HAAS | Heuristic Algorithm with Adaptation Strategy |
| HOL | Head-Of-the-Line |
| HV | High Voltage |
| IEEE | Institute of Electrical and Electronics Engineers |
| IFFT | Inverse Fast Fourier Transform |
| ITU | International Telecommunication Union |
| LAN | Local Area Network |
| LV | Low Voltage |
| MAC | Medium Access Control |
| MODEM | MOdulation and DEModulation |
| MSC | Message Sequence Chart |
| MV | Medium Voltage |
| NB-PLC | Narrowband Powerline Communications |
| OFDM | Orthogonal Frequency Division Multiplexing |
| OFDMA | Orthogonal Frequency Division Multiple Access |
| OPERA | Open PLC European Research Alliance |

| | |
|---|---|
| PER | Packet Error Ratio |
| PHY | Physical |
| PLC | Powerline Communications |
| PMF | Probability Mass Function |
| PRIME | PoweRline Intelligent Metering Evolution |
| QAM | Quadrature Amplitude Modulation |
| QoS | Quality of Service |
| REMPLI | Real-time Energy Management via Powerline and Internet |
| PSK | Phase Shift Keying |
| RP | Repeater |
| RR | Round Robin |
| RRS | Round Robin Based Resource Distribution |
| RU | Resource Unit |
| SA | Simulated Annealing |
| SC | Street electricity Cabinet |
| SNR | Signal to Noise Ratio |
| SP | Service Provider |
| TDM | Time Division Multiplexing |
| TDMA | Time Division Multiple Access |
| TDR | Time Division Repeater |
| TF | Time Frame |
| ThU | Throughput Unit |
| UDP | User Datagram Protocol |
| UG*i* | User group *i* |
| UMTS | Universal Mobile Telecommunications System |
| UPA | Universal Powerline Association |
| VDSL | Very high speed Digital Subscriber Line |
| VoIP | Voice over Internet Protocol |
| VPN | Virtual Private Network |

| | |
|---|---|
| WAN | Wide Area Network |

## A. 5 List of Symbols

| | |
|---|---|
| $\alpha$, $\beta$ | Scale parameters |
| $a_{i,f}$ | Variable indicate if frequency $f$ is assigned in cell $i$ (1) or not (0) |
| $\boldsymbol{A}_{CA}$ | Channel allocation matrix |
| $B_{Ch}$ | Total number of allocated channels in network |
| $B_{Ci}$ | Number of channels allocated in cell $C_i$ |
| $B_{Ei}$ | Effective number of channels in cell $C_i$ |
| $B^-_{C_i}$ | Number of channels to be released by $C_i$ |
| $B_{Ui}$ | Benefit from a user $U_i$ |
| $B_U$ | Benefit of a scheduling solution |
| $BS_i$ | Base station of cell $C_i$ |
| $\boldsymbol{C}$ | Set of cells |
| $\boldsymbol{C}^{(j)}_{nb}$ | Set of neighboring cells of cell $C_j$ |
| $D_{j\to i}$ | Duration for delivery a channel announcement message (CA-Msg) from $C_j$ to $C_i$ |
| $\overline{D_{\to i}}$ | Average delivery time of CA-Msgs to cell $C_i$ |
| $\overline{D_{Ch}}$ | Average delivery time of CA-Msgs in the PLC network |
| $\boldsymbol{F}$ | Set of channels |
| $F$ | Number of channels |
| $f_j$ | Channel with index $j$ |
| $F_{Ch}$ | Inter cell fairness index |
| $F_U$ | Intra cell fairness index |
| $f_{CR}$ | Channel reuse factor |
| $G$ | Fitness of a scheduling solution |
| $\Delta G^+_{i,j}$ | Fitness gained by assigning resources to service $q_j$ of user $U_i$ |
| $\boldsymbol{I}_{in}$ | Interference matrix |

| | |
|---|---|
| $I_{Ch}$ | Number of interferences in the whole PLC network |
| $i,j,t,k$ | Variables |
| $i_{i,j}$ | Interference variable between cell $C_i$ and $C_j$ |
| $L$ | Packet length |
| $M_i$ | Number of allocated channels per active user |
| $M_{NB}$ | $M$ value of a neighboring PLC cell |
| $\mathbb{N}$ | Natural numbers |
| $N_i$ | Number of users in PLC cell $i$ |
| $\boldsymbol{N_U}$ | Set of users in each cell in the PLC network |
| $N^{Carrier}$ | Number of carriers |
| $n^{conn}$ | Number of logical connections between PLC cells |
| $p_j^{send}$ | Probability of sending a CA-Msg from cell $C_j$ in a time frame |
| $p_{j\rightarrow i}^{recv}$ | Probability of receiving a CA-Msg from cell $C_j$ in $C_i$ in a time frame |
| $p_i^{recv}$ | Probability of receiving a CA-Msg in $C_i$ in a time frame |
| $p_{on}$ | Transition probability from Off to On |
| $p_{off}$ | Transition probability from On to Off |
| $P_{serv}^{(k)}$ | Probability that service $k$ is available in a time frame |
| $P_{no\_serv}^{(k)}$ | Probability that service $k$ is not available in a time frame |
| $P_{serv}^{(U_i)}$ | Probability that there is at least one service of user $U_i$ in a time frame |
| $P_{no\_serv}^{(U_i)}$ | Probability that there is no service of user $U_i$ in a time frame |
| $Q$ | Number of services |
| $q_j$ | Service $j$ |
| $R_{i,j}$ | Resources for service $q_j$ of user $U_i$ |
| $R^{OH}$ | Resources for overhead of transmission |
| $R_{Ui}$ | Resources for user $U_i$ |
| $RU$ | Resource Unit |

| | |
|---|---|
| $S^{PHY}$ | Data rate in the physical layer |
| $S$ | Number of cells |
| $S^{Th}$ | Throughput |
| $T_{sym}$ | Duration of a OFDM symbol |
| $T_{TF}$ | Duration of a time frame |
| $T_{j \to i}^{CA\_Msg}$ | The time that a CA-Msg from $C_j$ is received by $C_i$ |
| $T_i^{ex}$ | Expiration time of a CA-Msg received from cell $C_j$ |
| $T_{i,j}$ | Transmission time for service $q_j$ of user $U_i$ |
| $T_{i,j,f}$ | Transmission time for service $q_j$ of user $U_i$ in channel $f$ |
| $T^A$ | Duration of access medium |
| $\mathbf{T}$ | Transmission time allocation matrix |
| $\boldsymbol{U}_i$ | Set of users in PLC cell $i$ |
| $U_i$ | Index of user number $i$ in set U |
| $\mathbf{X}$ | Service grant matrix |
| $x_{i,j}$ | Number of granted services $j$ of user $i$ |
| $\mathbf{Y}$ | Service request matrix |
| $y_{i,j}$ | Number of requested service $j$ of user $i$ |

# Curriculum Vitae

Name: Le Phu Do

Date of birth: 02.02.1977, in Hanoi, Vietnam

Address: Hochschulstr. 34/303, 01069-Dresden, Germany

E-Mail: dolephu@ifn.et.tu-dresden.de

## Education and Professional Experience

| | |
|---|---|
| 09/1982 - 05/1994 | School in Bacninh, Vietnam |
| 10/1994 - 05/1999 | Bachelor degree in Telecommunications, Hanoi University of Technology, Vietnam |
| 10/1999 – 10/2001 | Lecture assistant at the Faculty of Electronics and Telecommunications, Ha Noi University of Technology, Vietnam |
| 10/2001 - 03/2004 | Master of Science degree in Telecommunications, Technische Universität Dresden, Germany |
| Since 08/2004 | Research assistant at the Chair for Telecommunications, Communications Laboratory, Technische Universität Dresden, Germany<br><br>PhD Thesis: *Hierarchical Resource Allocation for Powerline Communications (PLC) Networks*, date of submission: 02.04.2012, date of defense: 18.09.2012 |